William Curtis

A catalogue of the British, medicinal, culinary and agricultural plants

Cultivated in the London botanic garden

William Curtis

A catalogue of the British, medicinal, culinary and agricultural plants
Cultivated in the London botanic garden

ISBN/EAN: 9783741139819

Manufactured in Europe, USA, Canada, Australia, Japa

Cover: Foto ©berggeist007 / pixelio.de

Manufactured and distributed by brebook publishing software
(www.brebook.com)

William Curtis

A catalogue of the British, medicinal, culinary and agricultural plants

A

C A T A L O G U E

OF THE

BRITISH, MEDICINAL, CULINARY,
AND AGRICULTURAL

P L A N T S,

CULTIVATED IN THE

LONDON BOTANIC GARDEN.

By W I L L I A M C U R T I S,
Author of the FLORA LONDINENSIS.

To which are prefixed,

P   R   O   P   O   S   A   L   S

For opening it by Subfcription.

L O N D O N:
Sold by B. WHITE, Fleet-ftreet; SEWEL, Cornhill;
ROBINSON, Pater-nofter Row; PAYNE, Pall-Mall; and
DEBRETT, Piccadilly.
M DCC LXXXIII.

[Price Three Shillings and Six-pence.]

TO THE HONOURABLE

DAINES BARRINGTON,

THOMAS WHITE, Esq.

AND THE REST OF THE

SUBSCRIBERS

TO THE

LONDON BOTANIC GARDEN,

THIS

CATALOGUE OF ITS CONTENTS,

IS WITH THE SINCEREST

GRATITUDE AND RESPECT,

INSCRIBED BY

W. CURTIS.

LIST of the SUBSCRIBERS.

ROGER Altham, Efq.
 Mr. William Alexander.
Right Honourable the Earl of Bute.
Honourable Daines Barrington.
Sir Jofeph Banks, Baronet.
John Baker, Efq.
Mrs. Rachael Barclay.
George Buxton, M. D·
Peter Calvert, LL. D.
Mr. Richard Calvert
Mr. John Dyer.
John Delamare, Efq.
Mr. John Darby
H. Grimfton, Efq.
Meffieurs Gordon, and Co.
Nich. Gwyn, M. D.
Bufick Harwood, .M. D.
Robert Hallifax, Efq.
Mrs. Hill
John Ibbettfon, Efq.

Robert

Robert Jenner, Efq.

Rev. Richard Kaye, D. D.

Mrs. King.

J. C. Lettfom, M. D.

George Morris, Med. Stud.

Right Honourable the Earl of Plymouth.

Mr. William Parker.

John Rawlinfon, M. D.

John Sims, M. D.

Rev. G. S. Townly

Mrs. Vafton.

James Vere, Efq.

Hon. Thomas Fr. Wenman.

Thomas White, Efq.

Dr. William Wynne

Mr. Thomas Willis

William Watfon, M. D.

Mr. Abraham Winterbottom.

Mr. William Wooton, Apothecary.

Mr. William Zachary

SITUATION.

THE *London Botanic Garden* is fituated very near the *Magdalen-Hofpital*, *St. George's-Fields*, in the road from the faid hofpital to *Weft-minfter-Bridge Turnpike*, through Lambeth Marfh village.

Its fituation being low, renders it peculiarly favourable to the growth of aquatic and bog plants, and all fuch as love a moift bottom, an ineftimable advantage in dry fummers.

TERMS OF SUBSCRIBING.

PERSONS fubfcribing one guinea a year are entitled to walk in the garden, ufe the library, and introduce one perfon.

Perfons fubfcribing two guineas a year have the additional privilege of introducing more than one, either in perfon or by ticket, and of receiving roots or feeds of fuch plants as can be fpared without diminifhing the neceffary ftock of the faid garden.

No flowers plucked, nor fpecimens taken but by permiffion.

The garden open to fubfcribers every day in the week, except Sunday.

Subfcriptions taken in at the Garden.

ERRATA.

Page 12, line 2, for *place* read place
—— 27, —— 2, — *curalavenfe* r. curaflavente
—— 27, —— 10, — *Broecklime* r. Brooklime
—— 31, —— 17, — *cineris* r. cinereus
—— 32, —— 11, — *Hofebound* r. Horehound
—— 32, —— 18, — *Oethufa* r. Acthufa
—— 35, —— 12, — *Saliva* r. Salvia
—— 65, —— 11, — *bouum* r. boum
—— 70, —— 19, — *Cannalis* r. Cannabis
—— 96, —— 29, — *Twayman* r. man
—— 96, —— 30, — *blade* r. Twayblade
—— 114, —— 19, — *Bur-red* r. Bur-reed

P R O P O S A L S, &c.

I T muſt be allowed, that all human knowledge ought to be ſubſervient to the good of ſociety, and in proportion as this is advanced by any ſcience, ſo ought that ſcience to be held in eſteem.

Providence in his unerring wiſdom, having allotted to mankind different capacities, and implanted in them propenſities to particular purſuits; ſo that what is matter of the greateſt ſatisfaction to one, ſhall be perfectly inſipid to another, it is no wonder that they ſhould differ ſo widely in the apprehended utility of their reſpective employments; each from a principle predominant in the human mind, being willing to think his own of the greateſt importance; and ſo ſtrongly is this idea impreſſed on the minds of ſome, as to betray them into a narrowneſs of thinking, inconſiſtent with that liberality of ſentiment, which would excite a wiſh for the univerſal increaſe of ſcience, when connected, even in the remoteſt degree, with the intereſts of mankind.

B   Without

Without derogating then in the least from the utility of other arts, I shall, in a few words, endeavour to point out the advantages of BOTANY, a science which this design is particularly intended to promote.

Among all the studies which engage mankind, few are attended with circumstances equally pleasing in their pursuit; few can boast that infinite variety of objects which are perpetually engaging our attention, and inviting us to partake of pleasures equally rational and innocent *.

It is a science which has been cultivated by the wiseft of mankind, and particularly by the most distinguished professors of the medical art.—Here, as philosophers, we may admire, and contemplate the beautiful works of an Almighty Being; what an infinite display of wisdom is observable in the different modes of the growth, and propagation of

* MILTON describes our first parents recreating themselves with a kindred employment.

" Awake, the morning shines, and the fresh field
" Calls us, we lose the prime, to mark how spring
" Our tender plants, how blows the citron grove,
" What drops the myrrh, and what the balmy reed."

plants !

plants! what care is taken in perpetuating the fucceffion of each fpecies! how admirably are they
adapted to grow in every different foil, and fituation, fo as to leave no part of the globe, not even
rocks and ftones uncovered! with what regular order, and in what conftant fucceffion, do they
flower and produce their feeds! In fhort, a perfon
of an inquifitive, or contemplative turn will find
in plants an endlefs fource of innocent delight;
another world, as it were, opens to his view, he
beholds the face of nature through a new medium
of vifion, and has the fuperior pleafure of being
able to read in that book which to the generality
of mankind is a mere blank.

The importance of this fcience as a branch of
medical knowledge is happily expreff'd by the late
ingenious Dr. Gregory, the ornament of his pro-
feffion, and I may add of human nature; in his
advice to the young Phyfician, he thus delivers his
fentiments on this head : " *The fcience of botany is*
" *fubfervient to the practice of phyfic as far as it facili-*
" *tates the knowledge of plants, by reducing them into*
" *the moft commodious and perfect fyftem ; and although*
" *it is not neceffary to be particularly acquainted with the*
" *name and hiftory of every plant, yet every one ought to*
" *be*

" *be so well founded in the principles of botany as to be*
" *able to find its place in the system, and to describe it*
" *scientifically:* AND WE OUGHT TO BE PARTICU-
" LARLY ACQUAINTED WITH EVERY MATE-
" RIAL CIRCUMSTANCE RELATIVE TO THOSE
" PLANTS WHICH ARE USED IN DIET OR ME-
" DICINE."

· It is however, much to be regretted, that this
science is so little attended to by gentlemen of the
faculty in this country, as it obliges them to de-
pend on the skill of the ignorant, and illiterate,
for many of their efficacious officinal plants, frequently at the expence of their own characters,
and of all that is valuable to their patients.

Although newly discovered chemical remedies
and foreign drugs, may have justly superseded many
of our English plants, yet a great number are still
retained in our Pharmacopœa, and many possess
very poisonous qualities; to be acquainted with
these at least, is the duty of every one, that takes
on himself the important character of guardian
of the healths of mankind.

. But it is not to physic alone, that botany is sub-
servient, perhaps it may be applied with as much
advantage

advantage to agriculture, as to any other science. In this enlightened age, when arts and sciences are carried to a pitch unthought of in former times, we might expect a nation celebrated not less for its arts, than its arms, would be the first to promote a science, whose improvements are the only solid check to the baneful, and enervating effects of luxury, and dissipation; and accordingly we find many of our nobility, gentlemen of landed property, and public societies, fully aware of its importance, and endeavouring by premiums, and a variety of other means, to improve it; much however still remains to be done, nor is it probable that their endeavours will be crowned with success, 'till botany is more cultivated, and plants, particularly the grasses, better understood.

How incapable most of our modern writers on agriculture are of communicating their discoveries for want of botanic information; and how much the progress of this most useful science is thereby retarded, must be obvious to all those who have perused their writings with any degree of attention.

I am inclined to suppose, that this inattention of the faculty and others to botany, proceeds in a great

great meafure from a want of opportunity to acquire it, and that if the means were afforded,
there would no longer be caufe of complaint or
cenfure.

To afford the means of obtaining this knowledge is the object of the prefent inftitution—an
inftitution which has been attended with con-
fiderable expence, and coft the author much time
and attention, fhould he have the pleafing fatis-
faction of feeing it become productive of national
utility, that time he fhall think ufefully employ-
ed and that attention moft happily beftowed.

The more effectually to fucceed in promoting
the knowledge previoufly recommended, as fo
effential to the interefts of individuals, and the
community at large, he has felected from the
boundlefs field of vegetable productions, certain
claffes of plants, univerfally acknowledged to be
either the moft ufeful, or the moft neceffary to be
known; by which means the ftudent's attention
is more immediately directed to the objects of his
purfuit, thefe are the *medicinal, culinary, poifonous,
agricultural* and *britifh* plants, all of which claffes
are kept in *feparate and diftinct quarters,* exprefs'd

in

in the plan of the garden ; to thefe are added two
other quarters, the one containing fuch plants as
are calculated to inftruct the ftudent in the principles of the Linnæan fyftem, being living ex-
amples of moft of his claffes and orders, the other
furnifhed with hardy, ornamental flowers and
fhrubs, chiefly exotic, and cultivated in the gardens of the curious.

As the practical part of Botany as well as of
every other fcience is the moft ufeful, fo it is pre-
fumed, the mode of communicating this knowledge
is fuch as will meet with general approbation, this
is effected by having the generic and trivial name
of each plant according to Linnæus, painted in a
legible hand, and affixed to it, and that none may
lofe the advantage of acquiring a knowledge of
plants from a non-acquaintaince with Latin, the
Englifh names alfo are added, with a view that
Botany in this familiar drefs, might be inftructive
to thofe, whom the bare mention of a long hard-
founding Latin name might tend to difcourage.

And the author is ready to flatter himfelf that
many perfons who are naturally fond of plants
and flowers, will be ready to encourage an undertaking

taking of this kind, by which, at the same time, that they indulge their particular taste, they may insensibly acquire knowledge

It has been objected by a few, that a knowledge of plants thus easily acquired, is as easily forgotten; but this must certainly be more the fault of the students, than the method, as they may without molestation spend as much time as they please in investigating them, and becoming perfectly acquainted with their characters; and, still farther, to assist them in their studies, a library of those books most necessary for students is open to their inspection.

It now remains for the author to express his gratitude to those who have patronized, encouraged, and assisted him in his undertaking.

To the generosity and public spirit of the honourable *Daines Barrington* and *Thomas White*, Esq. his principal patrons in this undertaking, the garden in a great degree owes its existence.

From his Majesty's matchless collection of plants in the *Royal Garden* at *Kew*, kept up with the most

unre-

unremitting attention by that intelligent botanist and gardener, *Mr. Aiton,* he has had the honour to receive many fcarce and valuable plants, both britifh and foreign; as alfo from the gardens of the *Earl of Bute,* at *Luton;* the *Duchefs Dowager of Portland,* at *Bulftrode Park;* the late *Dr. Fothergill,* at *Upton;* *Dr. Pitcairn,* at *Iflington;* *Dr. Lettfom,* at *Grovehill, Camberwell;* and the *Apothe-caries Company,* at *Chelfea.*

He is alfo indebted for plants to the Right Hon. the *Earl of Plymouth; Sir Jofeph Banks, Baronet; Rev. J. Lightfoot, Uxbridge; Rev. Dr. Goodenough, Ealing; Thomas Frankland, Efq; York; T. G. Cullum, Efq; Reverend Mr. Laurents, Bury; James Crowe, Efq; Norwich; Thomas Woodward, Efq; Bungay; Mr. Dickfon, Covent-Garden; Mr. Sole, Bath; Mr. William Fothergill, Wenfleydale, Yorkfhire; Rev. T. Woodford, Winchefter; Rev. J. Davies; Rev. R. Relhan, Cambridge; Dr. Calvert, LL D.; Mr. Grimfton, Northampton; Mr. Howard, Knutsford, Chefhire; Mr. Cockfield, Upton; Mr. Wagftaff, Norwich; Mr. Sparfhall, Yarmouth; Mr. Alexander, Hallifax; Mrs. Curtis, Alton; Mrs. King, Blackheath; Mr. Wheeler, Mr. Willis, Mr. Sibley, Mr. Ridout, Mr. Upham, London; Meffrs. Lee, Gordon, Malcolm, Drivers, Lod-*

C   *diges*

*diges and Grimwood, Nurferymen; Rev. Mr. Lyon.,
Dover; and Mr. Rayer, London*; he is fearful he
has omitted many other kind contributors, he
hopes fuch will not be offended, and, that there
are yet many perfons in different parts of Great
Britain who will not need the ftimulus of example
to furnish the author with many of his defiderata,

CON-

Pages 19-20 missing

A

CATALOGUE

OF THE

MEDICINAL PLANTS

CULTIVATED IN THE

LONDON BOTANIC GARDEN.

Medicus omnium ſtirpium (ſi fieri poteſt) Peritiam habeat :
Sin minus plurium ſaltem, quibus frequenter utimur.
GALEN. *Lib. I. Antidot.*

PROME etiam, feu tunde prius, feu contere gyro
    Quod viride hortus habet, vel quod carnaria ficcum,
Allia, ferpyllumque herbas, thymbramque falubrem,
Braſſicaque et raphanos, ac longis intyba fibris
Et mentam, et finapi, coriandrum, prototomumque,
Erucam, atque apium, malvam, betamque falubrem,
Rutamque et nafturcum, et amara abſinthia mifce,
Puleiumque potens, nec non et lene cyminum.
Palmula nec defint Idumes, nec pruna Damafci.
Adde et aromaticas fpecies, quas mittit Eous
Vel quæ Judaicis fragrant bene condita capſis
Tus, coftum, folium, myrrham, ſtyracem, crocomagma,
Afpalathum, gallam, elleborum, nigrumque bitumen,
Et nardum, et caſias, et amoma, et cinnama rara,
Balfama, peucedanum, fpicam, crocum atque bedellam,
Irim, caftoreum, fcillamque, opium, panaceam,
Reſinam, lepidum, euforbium, git atque pyretrum.
Zinziber et calidum, mordax piper, et lafer algens,
Agaricumque, afarumque potens, aloën, aconitum,
Galbana, fandaracam, famfucum, pforicum, alumen,
Acaciam, propolimque, et adarchen, cnicon, acanthum,
Andrachnen, acoronque, opopanaca, pompholygemque,
Cyperum, ladanum, fagapenonque, et tragacanthum,
Scammoniam, cypen, malabathron, ammoniacon.

Marcellus de Medicina.

B^Y the wife and unchangeable laws of nature, eftablifhed by a Being infinitely good, and infinitely powerful, not only Man, the lord of the creation, " fair form who wears fweet fmiles, and looks erect on heaven," but every fubordinate being, becomes fubject to decay and death; pain and difeafe, the inheritance of mortality, ufually accelerate his diffolution; to combat thefe, to alleviate when it has not the power to avert, *Medicine*, honoured art, comes to our affiftance.

It will not be expected that we fhould here give a hiftory of this ancient practice, or draw a parallel betwixt the fuccefs of former Phyficians and thofe of modern times; all that concerns us to remark is, that the ancients were infinitely more indebted to the vegetable kingdom for the materials of their art than the moderns : not fo well acquainted with the œconomy of nature, which teaches us that plants were chiefly deftined for the food of various animals, they fought in every herb fome latent healing virtue, and frequently endeavoured to make up the want of efficacy in one by the combination of numbers; hence the extreme length of their farraginous prefcriptions.

More

More enlightened ideas of the operations of medicines, joined to feveral powerful remedies drawn from the mineral kingdom, have taught the moderns greater fimplicity and concifenefs in practice; perhaps there is a danger that this fimplicity may be carried too far, and become finally detrimental to the exercife of the healing art.

In felecting the plants ufed, in the practice of phyfic, commonly called *officinals*, we have confined ourfelves to thofe enumerated in the difpenfatories of this kingdom, dividing them into two claffes, one of which may be diftinguifhed by the name of *ufitatæ*, and the other *minus ufitatæ*; in the former we include all thofe of the prefent LONDON and EDINBURGH *Pharmacopœias*, in the latter fuch as are omitted by the colleges, but retained in LEWIS's Difpenfatory, with fome few of modern introduction.

Thofe plants which have numbers prefixed to them, are at prefent growing in the garden, thofe which have none have not yet been obtained, and fuch as have an afterifk are either the produce of warmer climates, and to be found only in the ftoves and confervatories of the curious, or incapable of being cultivated.

MEDICINAL PLANTS

OF THE

LONDON AND EDINBURGH

DISPENSATORIES.

L. E.	1 Abrotanum mas	Southernwood	Artemisia Abrotanum
L.	2 Absinthium maritimum	Sea Wormwood	Artemisia maritima
L. E.	3 Absinthium vulgare	Common Wormwood	Artemisia Absinthium
L.	* Acacia	Acacia	Mimosa nilotica
E.	4 Acetosa	Sorrel	Rumex Acetosa
E.	5 Aconitum	Blue Monkhood	Aconitum Napellus
L.	* Agaricus	Larch Agaric	Boletus Pini Laricis
E.	* Agaricus	Oak Agaric	Boletus igniarius
L. E.	6 Allium	Garlic	Allium Sativum

Medicinal Plants of the London and Edinburgh Dispensatories.

L. E.	•	Aloe Socotorina	Socotrine Aloes	} Aloe perfoliata
L. E.	•	Aloe Hepatica	Horse Aloes	}
L. E.	7	Althæa	Marshmallow	Althæa officinalis
L.	8	Ammi	Common Bishop's Weed	Ammi majus
L.	9	Amomum	Common Amomum	Sison Amomum
L. E.		Ammoniacum	Gum Amoniac	Ammoniacum
L. E.	10	Amygdalæ { Amaræ / Dulces	Almonds { Bitter / Sweet	} Amygdalus communis
L. E.		Anchusa	Alkanet	Anchusa tinctoria
L. E.	11	Anethum	Dill	Anethum graveolens
L. E.	12	Angelica sativa	Garden Angelica	Angelica Archangelica
L. E.	13	Angelica sylvestris	Wild Angelica	Angelica sylvestris
L. E.	14	Anisum	Aniseed	Pimpinella Anisum
L. E.	15	Aristolochia tenuis	Slender Birthwort	Aristolochia Clematitis
L. E.		Aristolochia longa	Long Birthwort	Aristolochia longa
L. E.	16	Artemisia	Mugwort	Artemisia vulgaris
L. E.	17	Arum	Cuckowpint	Arum maculatum
L. E.	•	Assafœtida	Asafœtida	Ferula Assafœtida
L. E.	18	Asarum	Asarabacca	Asarum Europæum
L. E.	19	Atriplex olida	Stinking Orach	Chenopodium Vulvaria

Medicinal Plants of the London and Edinburgh Dispensatories.

L. E.	* Aurantium hifpalense	Seville Orange	⎰ Citrus Aurantium
L. E.	* Aurantium curalavense	Curaffoa Orange	⎱
L. E.	20 Balnuftia	Balauftines	Punica Granatum
L. E.	21 Balfamum canadenfe	Canada Balfam	Pinus balfamea
L. E.	* Balfamum Copaiva	Balfam Capivi	Copaifera officinalis
L. E.	* Balfamum Peruvianum	Balfam of Peru	Peruifera
E.	* Balfamum Tolutanum	Balfam of Tolu	Toluifera Balfamum
E.	22 Bardana	Burdock	Arctium Lappa
L.	* Bdellium	Bdellium	Bdellium
L.	23 Becabunga	Brooklime	Veronica Becabunga
L. E.	24 Belladonna	Deadly Nightfhade	Atropa Belladonna
L. E.	* Benzoinum	Gum Benjamin	Croton Benzoë
L. E.	25 Biftorti	Biftort	Polygonum Biftorta
L. E.	26 Bryonia	Bryony	Bryonia alba
	27 Buxus	Box	Buxus fempervireus
L. E.	28 Calamus aromaticus	Sweet Flag	Acorus Calamus
L.	29 Calamintha	Field Calamint	Meliffa Nepeta
L. E.	* Camphora	Camphor	Laurus Camphora
L. E.	* Canella alba	Falfe Winter's Bark	Winterania Canella
L. E.	* Cardamomum	Cardamoms	Amomum Cardamomum
L. E.	30 Carduus benedictus	Bleffed Thiftle	Centaura benedicta

Medicinal Plants of the London and Edinburgh Dispensatories.

L.	31	Carica	Fig	Ficus Carica
L.		Carpobalfamum	Carpobalfam	Amyris Opobalfamum
L. E.	32	Carum	Caraway	Carum Carui
L. E.	*	Caryophylla aromatica	Clove	Caryophyllus aromaticus
L. E.	33	Caryophylla rubra	Clove july flower	Dianthus Caryophyllus
L. E.		Cafcarilla	Cafcarilla	Croton Cafcarilla
L. E.	*	Caffia fiftularis	Caffia fiftula	Caffia fiftula
L.		Cafumunar	Cafumunar	
L. E.	34	Centaurium minus	Centaury	Gentiana Centaurium
L.	35	Cepa	Onion	Allium Cepa
L.	36	Chamædrys	Germander	Teucrium Chamædrys
L. E.	37	Chamæmelum	Camomile	Anthemis nobilis
L.	38	Chamæpitys	Groundpine	Teucrium Chamæpitys
L.	39	Cicuta	Hemlock	Conium maculatum
L. E.	*	Cinnamomum	Cinnamon	Laurus Cinnamomum
L. E.	40	Cochlearia	Scurvy-grafs	Cochlearia officinalis
L. E.	41	Colchicun	Meadow-faffron	Colchicum autumnale
L. E.	*	Colocynthis	Coloquintida	Cucumis Colocynthis
L. E.	*	Colomba	Columbo root	
L. E.	#	Contrayerva	Contrayerva	Dorftenia Contrayerva
E.	42	Convallaria	Solomon's Sed	Convallaria Polygonatum

Medicinal Plants of the London and Edinburgh Dispensatories.

L. E.	45	Coriandrum	Coriander	Coriandrum fativum
L. E.	*	Cortex peruvianus	Jefuits Bark	Cinchona officinalis
L.	*	Coftus	Coftus	Coftus arabicus
L.	E. 44	Crocus	Saffron	Crocus fativus
L.	*	Cubebæ	Cubebs	Piper caudatum
L.	45	Cucumis agreftis	Elaterium	Momordica Elaterium
L. E.	*	Curcuma	Turmeric	Curcuma longa
L. E.	46	Cydonia mala	Quinces	Pyrus Cydonia
L. E.		Cyminum	Cummin	Cuminum Cyminum
L.	47	Cynofbatos	Hips	Rofa canina
L.	*	Daucus creticus	Candy Carrot	Athamanta cretenfis
	E. 48	Daucus fylveftris	Wild Carrot	Daucus Carota
L.	49	Dictamnus creticus	Ditany of Crete	Origanum Dictamnus
	E. 50	Dulcamara	Woody Nightfhade	Solanum Dulcamara
L.	51	Elatine	Fluellin	Antirrhinum Elatine
L.		Eleutheria, *vid.* Cafcar.		
L.	E. 52	Enula campana	Elecampane	Inula Helenium
L.	E. 53	Eryngium	Eryngo	Eryngium maritimun
L.	E. 54	Flammula Jovis	Upright Travelle's Joy	Clematis recta
L.	E.	Fœniculum dulce	Sweet Fennel	} Anethum Fœniculum
	E. 55	Fœniculum vulgare	Fennel	

Medicinal Plants of the London and Edinburgh Difpenfatories.

L. E.	56	Fenum græcum	Fenugreek	Trigonella Fænum græcum
L. E.	57	Fumaria	Fumitory	Fumaria officinalis
L. E.	*	Galbanum	Galbanum	Bubon Galbanum
L. E.	*	Gambogia	Gamboge	Cambogia Gutta
L. E.	58	Genifta	Broom	Spartium Scoparium
L. E.	59	Gentiana	Gentian	Gentiana lutea
L. E.	60	Gladiolus luteus	Yellow-flag	Iris Pfeudacorus
L. E.	61	Glycyrrhiza	Liquorice	Glycyrrhiza glabra
L. E.	62	Granatum	Pomegranate	Punica Granatum
L. E.	**	Gujacum	Guajacum	Guajacum officinale
L. E.	* *	Gummi Ammoniacum	Gum Ammoniac	Ammoniacum
L. E.	*	Gummi Arabicum	Gum Arabic	Mimofa nilotica
L. E.	*	Gummi Elemi	Gum Elemi	Amyris elemifera
L. E.	63	Gummi Tragacantha	Gum Tragacanth	Aftragalus Tragacantha
L. E.	64	Hedera terreftris	Ground-ivy	Glechoma hederacea
L. E.	65	Helleborus albus	White Hellebore	Veratrum album
L. E.	66	Helleborus niger	Black Hellebore	Helleborus niger
L. E.	67	Hippocaftanum	Horfe Chefnut	Æfculus Hippocaftanum
L.	68	Hordeum diftichum	Barley	Hordeum vulgare
E.	69	Hydrolapathum	Water Dock	Rumex Hydrolapathum
E.	70	Hyofcyamus	Henbane	Hyofcyamus niger

Medicinal Plants of the London and Edinburgh Dispensatories.

L. E. 71	Hypericum	St. John's Wort	Hypericum perforatum
L. *	Hypocistis	Hypocistis	Cytinus Hypocistis
L. E. 72	Hyssopus	Hyssop	Hyssopus officinalis
L. E.	Jalapium	Jalap	Convolvulus Jalapa
E. 73	Imperatoria	Masterwort	Imperatoria Ostruthium
L. E. *	Ipecacuanha	Ipecacuanha	Viola Ipecacua
L. E. 74	Iris florentina	Florentine Orrice	Iris florentina
L. E. 75	Iris palustris	Yellow Water-flag	Iris Pseudacorus
L. E.	Juncus odoratus	Camels hay	Andropogon Schœnanthus
L. E. 76	Juniperus	Juniper	Juniperus communis
L. *	Kermes	Kermes grains	Quercus coccifera
L. E. *	Kino	Gum Kino	
L.	Labdanum	Gum Labdanum	Cistus creticus
L. 77	Lamium album	White Archangel	Lamium album
L. E. 79	Lavandula	Lavender	Lavandula Spica
L. E. 80	Laurus	Bay	Laurus nobilis
L. *	Lichen cineris terrestris	Ash cold ground Liver-wort	Lichen caninus
L. E. 81	Ligusticum	Lovage	Ligusticum Levisticum
L. *	Lignum Rhodium	Rhodium Wood	Genista canariensis
L. E. *	Lignum campechense	Logwood	Hæmatoxylum campechianum

Medicinal Plants of the London and Edinburgh Dispensatories.

L. E.	*	Limonia mala	Lemon	Citrus medica
L. E.	12	Linum	Linseed	Linum usitatissimum
L.	13	Lujula	Wood forrel	Oxalis Acetofella
L.	*	Macis	Mace	Myristica
L. E.	34	Majorana	Marjoram sweet	Origanum Majorana
L.	*	Malabathrum	Indian leaf	Laurus Caffia
L. E.	35	Malva	Mallow	Malva sylvestris
L. E.	36	Manna	Manna	Fraxinus Ornus
L.	37	Marum fyriacum	Syrian herb Maslick	Teucrium Marum
L.	38	Marum vulgare	Herb Maslick	Thymus maslichina
L.	39	Marrubium	Hofehound	Marrubium vulgare
L. E.	*	Maf'che	Gum Maslick	Pistachia Lentiscus
L.		Melampodium, v Helleb. nig.		
L.	30	Matricaria	Fererfew	Matricaria Parthenium
L.	31	Melissa	Balm	Melissa officinalis
L. E.	32	Mentha piperitis	Peppermint	Mentha Piperitis
L. E.	33	Mentha vulgaris	Spearmint	Mentha viridis
L.	34	Menyanthes	Buckbean	Menyanthes trifoliata
	E. 35	Meum athamanticum	Spignel	Oethufa Meum
	E. 36	Mezereon	Mezereon	Daphne Mezereon
	L. 37	Millefolium	Yarrow	Achillea Millefolium

Medicinal Plants of the London and Edinburgh Dispensatories.

L.		98	Morus	Mulberry	Morus nigra
L. E.		*	Myrrha	Myrrh	Myrrha
L.		99	Napus	Sweet Navew	Braffica Napus
L.			Nardus celtica	Celtic Spikenard	Valeriana celtica
L.		*	Nardus indica	Indian Spikenard	Andropogon Nardus⌐
L. E.		100	Nafturtium aquaticum	Water-crefs	Sifymbrium Nafturtium
L.		101	Nepeta	Cat-mint	Nepeta Cataria
L.		102	Nicotiana	Tobacco	Nicotiana Tabacum
L. E.		*	Nux mofchata	Nutmeg	Myriftica
L.		*	Olea fativa	Olive.	Olea Europæa
L. E.		*	Olibanun	Olibanum	Juniperus lycia
L. E.		*	Opium, *vid.* Papaver		
L. E.		*	Opoponax	Opoponax	Paftinaca Opoponax
L.		103	Origanum	Wild Marjoram	Origanum vulgare
L.		104	Pæonia	Piony	Pæonia officinalis
E.		105	Palma Chrifti, *vid.* Ricinus		
E.		*	Palma oleofa	Palm-oil	Palma oleofa
L. E.		166	Papaver album	Garden Poppy	Papaver fomniferum
L.		107	Papaver erraticum	Corn Poppy	Papaver Rhœas
L.		108	Paralyfis	Cowflip	Primula veris
L. E.		109	Parietaria	Pellitory of the wall	Parietaria officinalis

E

Medicinal Plants of the London and Edinburgh Dispensatories.

	No.			
L.	110	Pentaphyllum	Cinquefoil	Potentilla reptans
L. *		Petrofelinum macedonicum	Macedonian Parfly	Bubon macedonicum
L. E.	111	Petrofelinum vulgare	Common Parfly	Apium Petrofelinum
L. E. *		Pimenta	Jamaica Pepper	Myrtus Pimenta
L.	112	Pimpinella Saxifraga	Burnet Saxifrage	Pimpinella Saxifraga
L. *		Piper album	White Pepper	Piper nig. decort.
L. E.		Piper Jamaicenfe, v. Pimenta		
L. E.		Piper longum	Long Pepper	Piper longum
L. E. *		Piper nigrum	Black Pepper	Piper nigrum
L.	113	Pix arida	Common Pitch	Pinus fylvestris
L. E.	114	Pix burgundica	Burgundy Pitch	Pinus Abies
L.	115	Pix liquida	Tar	Pinus fylvestris
L. E.	116	Plantago	Plantain	Plantago latifolia
L.	117	Polium	Poley-mountain	Teucrium creticum
L.	118	Pruna gallica	Prunes	Prunus domeflica
L.	119	Pruna fylveftria	Sloes	Prunus communis
L. E.	120	Pulegium	Pennyroyal	Mentha Pulegium
L. E.	121	Pulfatilla nigricans	Meadow Pafque flower	Anemone pratenfis
L.		Pyrethrum	Pellitory of Spain	Anthemis Pyrethrum
L. E.	122	Quercus	Oak	Quercus Robur
L. E.	123	Raphanus rufticanus	Horfe-radifh	Cochlearia Armoracia

Medicinal Plants of the London and Edinburgh Dispensatories.

L. E.	124 Rhabarbarum	Rhubarb	Rheum palmatum
E.	125 Ricinus	Palma Christi	Ricinus communis
L. E.	Rhamnus, *vid.* Spin. Cerv.		
L. E.	126 Rorismarinus	Rosemary	Rosmarinus officinalis.
L. E.	127 Rosa damascena	Damask Rose	Rosa centifolia
L. E.	128 Rosa rubra	Red Rose	Rosa gallica
L. E.	129 Rubia tinctorum	Madder	Rubia tinctorum
L. E.	130 Rubus Idaeus	Rasberry	Rubus Idaeus
L. E.	131 Ruta	Rue	Ruta graveolens
L. E.	132 Sabina	Savine	Juniperus Sabina
L. E.	* Saccharum	Sugar	Saccharum officinarum
L. E.	* Sagapenum	Sagapenum	Sagapenum
L. E.	133 Salvia	Sage	Salvia officinalis
L. E.	134 Sambucus	Elder	Sambucus nigra
L. E.	* Sanguis Draconis	Dragons Blood	Calamus Rotang
L. E.	Santalum citrinum	Yellow Saunders	Santalum album
L. E.	Santalum rubrum	Red Saunders	Santalum rubrum
L. E.	* Santonicum	Wormseed	Artemisia judaica
L. E.	* Sarcocolla	Sarcocol	Penea mucronata
L. E.	Sarsaparilla	Sarsaparilla	Smilax Sarsaparilla.
L. E.	Sassafras	Sassafras	Laurus Sassafras

Medicinal Plants of the London and Edinburgh Dispensatories.

	E.	135	Satyrion	Salep	Orchis mascula
L. E.	136	Scammonium	Scammony	Convolvulus Scammonium
L. E.	137	Scilla	Squill	Scilla maritima
L. E.	138	Scolopendrium	Hart's Tongue	Asplenium Scolopendrium
L. E.	139	Scordium	Water Germander	Teucrium Scordium
L. E.	*	Seneka	Rattlesnake root	Polygala Senega
L. E.	*	Senna	Senna	Cassia Senna
L. E.		Serpentaria Virginiana	Virginian Snake Root	Aristolochia Serpentaria
L. E.		Serpyllum, *vid.* Thymus		
L.		Sedlia	Hartwort	Laserpitium Siler
E.	140	Sigillum Solomonis	Solomon's Seal	Convallaria Polygonatum
E.	*	Simarouba	Simarouba	Quassia dioica
L. E.	141	Sinapis	Mustard	Sinapis nigra
L. E.	142	Spina cervina	Buckthorn	Rhamnus catharticus
L.	143	Stoechas	French Lavender	Lavandula Stoechas
F.	144	Stramonium	Thorn Apple	Datura Stramonium
L. E.	*	Styrax	Storax	Styrax officinale
L. E.	145	Symphytum	Comfrey	Symphytum officinale
L. E.	*	Tamarindus	Tamarind	Tamarindus indica
L. E.	146	Tanacetum	Tansy	Tanacetum vulgare
L. E.	147	Taraxacum	Dandelion	Leontodon Taraxacum

Medicinal Plants of the London and Edinburgh Dispensatories.

L.	E.	148	Terebinthina veneta	Venice Turpentine	Pinus Larix
		149	Terebinthina argentoratensis	Straſburgh ditto	Pinus Picea
L.			Terebinthina Chia	Cypreſs ditto	Piſtacia Terebinthus
L.		150	Terebinthina communis	Common ditto	Pinus ſylveſtris
L.	E.	*	Terra Japonica	Japan Earth	Mimoſa Catechu
L.		151	Thlaſpis	Treacle Muſtard	Thlaſpi arvenſe
L.		152	Thus	Frankincenſe	Pinus ſylveſtris
					{ Thymus vulgaris E.
L.	E.	153	Thymus	Thyme	{ Thymus ſerpyllum L.
L.		154	Tilia	Lime-tree	Tilia europæa
L.	E.	155	Tormentilla	Tormentil	Tormentilla erecta
L.	E.	156	Tragacantha	Gum Tragacanth	Aſtragalus Tragacantha
L.	E.	157	Trichomanes	Maiden hair	Aſplenium Trichomanes
L.	E.	158	Trifolium paludoſum	Buckbean	Menyanthes trifoliata
L.		159	Triticum	Wheat	Triticum hybernum
	E.	160	Tuſſilago	Coltsfoot	Tuſſilago Farfara
L.	E.	161	Valeriana	Valerian wild	Valeriana officinalis
L.	E.	162	Vitis	Vine	Vitis vinifera
			Veratrum, viz. Helleb. alb.		
	E.	163	Verbaſcum	Mullein tall	Verbaſcum Thapſus
L.	E.	164	Viola	Violet ſweet	Viola odorata

38

Medicinal Plants of the London and Edinburgh Dispensatories.

L.	165	Uvæ passæ majores	Raisins	Vitis vinifera
E.	166	Urtica	Nettle	Urtica dioica
E.	167	Uva Urfi	Bearberry	Arbutus Uva Urfi
L. E.	168	Zedoaria	Zedoary	Kæmpferia rotunda
L. E.	*	Zingiber	Ginger	Amomum Zingiber

O M I T T E D.

E.	*	Aurantium curaflavense	Curaflao Oranges	Citrus Aurantium
L. E.	169	Gallæ	Galls	Querς Cerris
E.	168	Refina alba	White Refin	Pinus Abies
L.	169	Refina flava	Yellow Refin	}
L.	*	Opobalfamum	Balfam of Gilead	Amyris Opobalfamum

MEDICINAL PLANTS

NOT CONTAINED IN THE

LONDON AND EDINBURGH

DISPENSATORIES.

1	Abies	Fir	Pinus Abies
2	Abrotanum fœmina	Lavender-cotton.	Santolina Chamæcyparissus
3	Absinthium ponticum	Roman Wormwood	Artemisia pontica
4	Acanthus	Bear's breech	Acanthus mollis
5*	Adianthum verum	True Maidenhair	Adianthum Capillus veneris
6	Agallochum	Aloes Wood	Agallochum
7	Ageratum	Maudlin	Achillea Ageratum
7	Agnus castus	Chaste tree	Vitex Agnus castus
8	Agrimonia	Agrimony	Agrimonia Eupatoria

Medicinal Plants not contained in the London and Edinburgh Dispensatories.

No.		English	
9	Alcea	Vervain Mallow	Malva Alcea
10	Alchemilla	Lady's Mantle	Alchemilla vulgaris
11	Alkekengi	Winter-cherry	Physalis Alkekengi
12	Alliaria	Jack by the Hedge	Erysimum Alliaria
*	Aloes caballina	Horse Aloes	Aloe perfoliata
13	Alnus vulgaris	Common Alder	Betula Alnus
14	Alnus nigra	Berry bearing Alder	Rhamnus Frangula
	Ammi verum	True Bishop's Weed	Sison Ammi
*	Amomum verum	True Amomum	
*	Anacardium orientale	Malacca Bean	Avicennia tomentosa
15	Anagallis	Pimpernel	Anagallis arvensis
*	Anime	Gum Anime	Hymenæa Courbaril
16	Anthora	Wholesome Wolf's Bane	Aconitum Anthora
17	Aparine	Goose Grass	Galium Aparine
18	Aquilegia	Columbine	Aquilegia vulgaris
19	Argentina	Silverweed	Potentilla Anserina
20	Aristolochia rotunda	Round Birthwort	Aristolochia rotunda
21	Arthanita	Sow Bread	Cyclamen europæum
22	Asperula	Woodroof	Asperula odorata
	Auricula Judæ	Jew's Ears	Peziza Auricula
23	Auricula Muris	Mouse Ear	Hieracium Pilosella

Pages 41-56 missing

55	Horse-radish	Cochlearia Armoracia
56	Jerusalem Artichoke	Helianthus tuberosus
57	Leeks	Allium Porrum
	Lemon	Citrus medica
58	Lentil	Ervum Lens
59	Lettuce	Lactuca sativa
	Limes	Citrus Aurantium
	Mace	Myristica
60	Marigold	Calendula officinalis
61	Marjoram pot	Origanum Onites
62	———— sweet	———— Majorana
63	Medlar	Mespilus Germanica
64	Melon	Cucumis Melo
	Mercury English, *vid.*	Good King Henry
	Pepper black	Piper nigrum
	———— white	——— decorticat.
	———— Jamaica	Myrtus Pimenta
	———— Cayenne	Capsicum annuum
	Pine-apple	Bromelia Ananas
	Pine-kernels	Pinus Pinea
	Pistachio Nut	Pistacia vera
65	Plumbs	Prunus domestica
66	Pomegranate	Punica Granatum
67	Potatoe	Solanum tuberosum
68	Prunes	Prunus domestica
69	Purslane	Portulaca oleracea
70	Quince	Pyrus Cydonia
71	Radish common	Raphanus sativus
	———— turnip rooted	—var. rad. rotund.
	———— Spanish	—var. rad. nigric.
72	Rampion	Campanula Rapunculus
73	Raisins	Vitis vinifera
74	Raspberry	Rubus idæus
	Rice	Oryza sativa
75	Millet	Panicum miliaceum
76	Monk's Rhubarb	Rheum Rhaponticum
	Morel	Phallus esculentus
77	Mulberry	Morus nigra
	Mushroom	Agaricus campestris

H

78	Muftard	Sinapis nigra
79	Nafturtium —————	Tropæolum majus
80	Nectarine	Amygdalus Perfica var. ··
	Nutmeg	Myriftica
81	Oat	Avena
	Olives	Olea Europæa
82	Onion common	Allium Cepa
83	—— Welch	——— fiftulofum
84	Orach garden	Atriplex fativa
	Orange	Citrus Aurantium
85	Parfley	Petrofelinum vulgare
86	Parfnep	Paftinaca fativa
87	Pea	Pifum fativum
88	Pear	Pyrus communis
89	Peach	Amygdalus Perfica
90	Rocambole ——	Allium Scorodoprafum
91	Rye	Secale cereale
92	Sage	Salvia
	Sago	Palma farinaria *Rumph.*
93	Salep	Orchis Morio
94	Salfafy	Tragopogon porrifolium
95	Savory Summer	Satureja hortenfis
96	—— Winter	——— montana
97	Samphire	Crithmum maritimum
98	Scorzonera ——	corzonera
	Sea-weed fweet ———	Fucus palmatus
	—— eatable —	—— efculentus
99	Service	Sorbus domeftica
100	Skirret	Sium Sifarum
101	Sloe	Prunus fylveftris
102	Spearmint	Mentha viridis
103	Spinach	Spinachia oleracea
104	Sorrel common	Rumex Acetofa
105	—— French	——— fcutatus
106	Strawberry	Fragaria fterilis
	Sugar	Saccharum officinarum
107	Tanfy	Tanacetum vulgare
108	Tarragon	Artemifia Dracunculus
	Tea Green	Thea viridis

	Tea Bohea	Thea Bohea
109	Thyme	Thymus vulgaris
	Truffles	Lycoperdon Tuber
110	Turnep	Brassica Rapa
111	Walnut	Juglans regia
112	Watercress	Sisymbrium Nasturtium
113	Wheat	Triticum hybernum
114	Whortleberry	Vaccinium Myrtillus

A

CATALOGUE

OF

POISONOUS PLANTS,

CULTIVATED IN THE

LONDON BOTANIC GARDEN.

" In the day that thou eateſt thereof thou ſhalt ſurely die."
Geneſis.

HOWEVER plaufibly the medical Practitioner may excufe his ignorance of plants in gene-ral, the public has a right to expect, that he fhould lofe no opportunity of making himfelf acquainted with thefe of the prefent clafs, fince they are few in number, and capable of producing in a fhort time the moft deleterious effects.

POISONOUS PLANTS.

1	Aconitum Napellus	Blue Monkſhood
2	Actæa ſpicata	Bane-berries
3	Atropa Belladonna	Deadly Nightſhade
4	Æthuſa Cynapium	Fool's Parſley
5	Cicuta viroſa *	Long leav'd WaterHemlock
6	Conium maculatum	Hemlock
7	Datura Stramonium	Thorn-apple
8	Hyoſcyamus niger	Henbane
9	Mercurialis perennis	Dog's Mercury
10	Oenanthe crocata	Hemlock Dropwort
11	Prunus Lauro ceraſus	Laurel
12	Solanum nigrum	Garden Nightſhade
13	—— Dulcamara	Woody Nightſhade

* Perſons not converſant in practical Botany, are very apt to confound this plant with the *Phellandrium aquaticum*, called alſo *Water-Hemlock*, an inſtance of this occurs in Mʀ. Wɪʟᴍᴇʀ's Pamphlet on poiſonous Vegetables, publiſhed in 1781, when Captain *Donellan's* trial was ſo generally the ſubject of converſation.

A

CATALOGUE

OF

AGRICULTURAL PLANTS,

Or such as are both useful and noxious

TO THE HUSBANDMAN,

CULTIVATED IN THE

LONDON BOTANIC GARDEN.

Agricola incurvo terram dimovit aratro.
Hinc anni labor : hinc patriam, parvofque nepotes
Suftinet : hinc armenta bouum, meritofque juvencosz
Nec requies, quin aut pomis exuberet annus,
Aut fœtu pecorum, aut Cerealis mergite culmi :
Proventuque oneret fulcos, atque horrea vincat.

Virgil. Georg. Lib. 2.

IT is presumed the present Catalogue is the first of the kind ever offered to the public; it will be no wonder, therefore, if it should fall short of perfection: most of the plants contained in it are of British growth; of the *useful* ones, many have long been known and cultivated in this country; several new ones are inserted, as deserving to be more generally known to the husbandman; of these the grasses, a much neglected tribe, form no inconsiderable part; at present, out of a hundred and three species, the produce of this country, considered by surrounding nations as most fertile in its herbage, only one is cultivated for pasturage, and that confessedly deficient in many of the requisites of a good grass: Mr. STILLINGFLEET, in his Miscellaneous Essays, has taken much pains to recommend several others as superior, and about as many more are here added from my own observations.

The Willows, and Poplars, though out of the herbaceous line, are included, being highly useful in rural œconomy, and but imperfectly known.

If

If the Agriculturift has been inattentive to the plants of the ufeful kind, the noxious plants have engaged ftill lefs of his attention, though highly deferving of it.

Future obfervations will doubtlefs make it neceffary to add many other plants to thofe here enumerated, perhaps retrench fome of the prefent lift, and throw a much greater light on the whole.

A

CATALOGUE

OF THE

PLANTS

USEFUL IN

AGRICULTURE.

PART I.

Barley common	Hordeum vulgare
—— big	—— hexastichon
—— sprat	—— distichon
—— long-eared	—— Zeocriton
Black-seed	Medicago lupulina
Buck-wheat	Polygonum Fagopyrum
Burnet great	Sanguiforba officinalis
—— small	Poterium Sanguiforba
Caraway	Carum Carui
Clover Dutch	Trifolium repens
—— broad-leaved	—— pratense
—— alpine, or narrow-leaved	—— alpestre
—— strawberry	—— fragiferum
Claver	Medicago Arabica
Coriander	Coriandrum sativum
Corn Indian	Zea Mays
—— Spelt	*Vid.* Wheat Spelt
Dyer's-weed	*Vid.* Weld
Flax	Linum usitatissimum
French-honeysuckle	Hedysarum coronarium
Grass Timothy	Phleum pratense

Grafs fweet-fcented Vernal	Anthoxanthum odoratum
—— meadow Foxtail	Alopecurus pratenfis
—— fine Bent	Agroftis capillaris
—— fmooth-ftalked Meadow	Poa pratenfis
—— rough-ftalked Meadow	—— trivialis
—— dwarf Meadow, or	
—— Suffolk grafs	—— annua
—— foft Brome	Bromus mollis
—— meadow Oat	Avena pratenfis
—— rough Oat	—— pubefcens
—— yellow Oat	—— flavefcens
—— fheep's Fefcue	Feftuca ovina
—— hard Fefcue	—— duriufcula
—— meadow Fefcue	—— pratenfis
—— fpiked Fefcue	—— pinnata
—— meadow Barley	Hordeum pratenfe
—— crefted Dog's tail	Cynofurus criftatus
—— Rye, or Ray	Lolium perenne
—— Canary	Phalaris canarienfis
Hemp	Cannalis fativa
Horfe bean	Vicia Faba
Indian Corn	Zea Mays.
Lucern	Medicago fativa
Madder	Rubia tinctoria
Maw feed	Papaver fomniferum
Muftard black	Sinapis nigra
—— white	—— alba
Nonefuch	Medicago Lupulina
Oat common	—— fativa
—— tartarian	— var. flofculis fecundis
—— naked	— nuda
Poplar black	Populus nigra
—— Lombardy	— var. ramis erectis
—— trembling	—— tremula
—— white	—— alba
—— Arbeal	—— Arbeel
Rib-grafs	Plantago lanceolata
Rape	Braffica Napus
Rye	Secale cereale

Saffron	Crocus officinalis
Saint-foin	Hedyfarum Onobrychis
Tare	Vicia fativa
Teafel	Dipfacus fullonum
Trefoil hop	Trifolium agrarium
—— procumbent	—— procumbens
—— bird's foot	Lotus corniculatus
Turnep	Braffica Rapa
Vetch common	Vicia fativa
Vetchling Meadow	Lathyrus pratenfis
Weld	Refeda luteola
Wheat common	Triticum hybernum
—— Poland	—— polonicum
—— coned	—— quadratum
—— Spelt	—— Spelta
Willow white	Salix alba
—— crack	—— fragilis
—— almond	—— amygdalina
—— yellow-barked	—— vitellina
—— fpurge-leaved	—— helix
—— fweet	—— pentandra
—— Ofier	—— viminalis
—— Sallow	—— caprea
Woad	Ifatis tinctoria

A

CATALOGUE

OF THE

PLANTS

NOXIOUS IN

AGRICULTURE.

PART II.

Bindweed fmall	Convolvulus arvenfis
———— great	———— fepium
Butterbur	Tuffilago Petafites
Blue-bottle	Centaurea Cyanus
Burdock	Arctium Lappa
Blite white	Chenopodium album . .
Brakes	Pteris aquilina
Biftort	Polygonum Biftorta
Buglofs fmall	Lycopfis arvenfis
Cockle	Agroftema Githago
Clown's Woundwort	Stachys paluftris
Cammock	Ononis arvenfis
Charlock	Sinapis arvenfis
Colt's foot	Tuffilago Farfara
Corn-marigold	Chryfanthemum fegetum
Crowfoot creeping	Ranunculus repens
——— round rooted	———— bulbofus
——— tall	———— acris
——— corn	———— arvenfis
Camomile corn	Matricaria Chamomilla
———— weak-fcented	———— inodora

Dock curled	Rumex crifpus
—— broad-leaved	—— obtufifolia
Fleabane common	Inula dyfenterica
Gout, or Afh-weed	Ægopodium Podagraria
Garlic crow	Allium vineale
—— bears	—— urfinum
Grafs garden Couch	Triticum repens
—— couchy Bent	Agroflis ftolonifera
—— couchy Oat	Avena elatior
—— bearded Oat	—— fatua
—— Lob	Bromus fecalinus
—— field Foxtail	Alopecurus agreflis
—— Darnel	Lolium temulentum
—— rough Cock's foot	Dactylis glomeratus
—— turfy Hair	Aira cefpitofa
—— meadow Soft	Holcus lanatus
—— creeping Soft	—— mollis
—— Carnation	Carex cefpitofa
Horfetail corn	Equifetum arvenfe
Hard-head's. *Vid.*	Knap-weed common
Knap-weed common	Centaurea nigra
——— great	—— Scabiofa
Mugwort	Artemifia vulgaris
Melilot	Trifolium Melilotus
Meadow-fweet	Spiræa Ulmaria
Muftard white	Sinapis alba
May-weed ftinking	Anthemis Cotula
Orach wild	Atriplex haftata
—— fpreading	—— patula
Oxeye Daify	ChryfanthemumLeucanthemum
Perficaria willow-leaved	Polygonum amphibium
——— fpotted-leaved	——— Perficaria
——— pale-flowered	——— Penfylvanicum
Poppy long fmooth-headed	Papaver dubium
—— round fmooth-headed	—— Rhæas
Radifh wild	Raphanus Raphaniftrum
Ramfons. *Vid.*	Bear's Garlick
Ragwort common	Senecio Jacobea
——— marfh	—— aquaticus

K

Reftharrow.	*Vid.*	Cammock
Rufh common		Juncus conglomeratus
—— blueifh		—— glaucus
—— flat-jointed		—— compreffus
—— round-jointed		—— articularus
—— bulbous		—— bulbofus
Succory blue		Cichorium Intybus
Sow-thiftle corn		Sonchus arvenfis
Spearwort fmall		Ranunculus Flammula
Scabious common		Scabiofa arvenfis
Spatling-poppy		Cucubalus Behen
Stinging-Nettle large		Urtica dioica
Silver-weed		Potentilla Anferina
Sneefe-wort		Achillea Ptarmica
Thiftle-milk		Carduus marianus
—— marfh		—— paluftris
—— melancholy		—— heienioides
—— fpear		—— lanceolatus
—— curfed		—— arvenfis
—— ftar		Centaurea Calcitrapa
Tine-Tare fmooth-podded		Ervum tetrafpermum
—— — rough-podded		—— hirfutum
Water horehound		Lycopus europæus
White back.	*Vid.*	Melancholy Thiftle
Yarrow		Achillea Millefolium
Yew		Taxus baccata

A

CATALOGUE

OF

BRITISH PLANTS,

CULTIVATED IN THE

LONDON BOTANIC GARDEN.

Arranged according to their

PERIODS OF FLOWERING.

The fall of kings,
The rage of nations, and the crush of states,
Move not the man, who from the world escap'd,
In still retreats, and flowery solitudes,
To Nature's voice attends, from month to month,
And day to day, thro' the revolving year;
Admiring sees her in her every shape,
Feels all her sweet emotions at his heart,
Takes what she lib'ral gives, nor thinks of more.

Thomson.

TO the diſtinguiſhed abilities, unwearied aſſiduity, and communicative diſpoſition of Mr. RAY, aſſiſted by many of his cotemporaries, we are indebted for the firſt *Britiſh Flora*, which could in any wiſe be conſidered as compleat; ſuch was his *Synopſis Stirpium Britannicarum*, the third Edition of which ſtill remains to be moſt deſervedly valued; in this work we have an enumeration, and frequently a deſcription, not only of the more perfect plants, but alſo of the *Muſci, Fuci, Fungi*, &c. all arranged according to a ſyſtem of his own. —To this ſyſtem of our great predeceſſor, ſucceeded that of the celebrated LINNÆUS, the ornament of Sweden; the excellence of which was ſuch as to create general admiration, and inſure it an almoſt univerſal reception: hence it became neceſſary (if I may be allowed the expreſſion) to *Linneanize* MR. RAY; this arduous taſk was attempted by MR. HUDSON, and MR. RAY's *Synopſis*, to ſpeak metaphorically, tranſmigrated into MR. HUDSON's *Flora Anglica*; in which form, though it retained the advantages of a new body, it loſt a portion of
its

its original fpirit, much ufeful information being fuppreffed, and many diftinct fpecies made into va-. rieties; it was, however, enriched with many new plants, difcovered by Mr. Hudson and his friends.

In the year 1777, Mr. Lightfoot having accompanied Mr. Pennant into Scotland, for the purpofe of making difcoveries in natural hiftory, favoured the public, on his return, with an ample account of the plants difcovered in his northern tour, under the title of *Flora Scotica*; the value of which was greatly enhanced by the new lights thrown on the cryptogamous plants in general, and the excellent plates of feveral rare plants which accompanied it.

In the fucceeding year appeared the fecond edition of Mr. Hudson's book, in which the plants of the *Flora Scotica*, with many additional ones previoufly difcovered, were inferted.

The works of thefe three different Authors, the only practical ones of any note that have been publifhed in this country, form the bafis of the prefent Catalogue; where *I have not* had an opportunity
of

of obferving or cultivating the plants which they
enumerate, I have relied implicitly on their accu-
racy; where *I have*, as an enquirer after truth, I
prefume I fhould have been cenfurable not to have
ufed the lights, which perhaps a more minute en-
quiry, or better opportunities of cultivation had
afforded me; hence I have in feveral inftances dif-
fered, both from Mr. RAY and Mr. HUDSON,
and am forry to have occafion to differ fo mate-
rially from the latter, in the account he has given
of the graffes, in the fecond edition of his work;
the reafons at large for fuch differences will be
given in the *Flora Londinenfis*, (vid. *Ranunculus hir-
futus, Polygonum minus, Sedum fexangulare*, &c.) the
only notice taken of them here will be to print in
Italics thofe plants which I apprehend to be fpecies,
and which Mr. HUDSON confiders as varieties.
As the Garden contains more than two-thirds of the
more perfect Britifh plants, it was apprehended that
the addition of the remainder, which either cannot
be made to grow, or have not yet been obtained,
would render this publication more ufeful, not
only to thofe who may have opportunities of vifit-
ing it, but to thofe alfo who have only the garden
which nature affords them; they are incorporated
with the reft, and diftinguifhed by having no fi-
gures

gures prefixed to them; such as have an asterisk are considered as doubtful natives.

As we have no compleat *Calendar* of a *British Flora*, it was thought that the utility of this little volume would be still farther enhanced, by arrang-ing the plants according to the months in which they usually flower in the garden; by this means, not only subscribers know what plants they have a right to expect in blossom at particular seasons, but the botanist at a distance may judge what plants he is likely to find in his herborizing excursions: allowances will naturally be made for variations of flowering, dependent on extraordinay seasons.

There is a circumstance relative to the blowing of certain plants, which deserves some attention here, I mean such as the *Groundsel, Shepherd's purse, Daisy, dwarf Meadow grass*, &c. I would observe, that though these may, if the winter prove mild, be found in blossom throughout the year, yet there is a certain month in which they flower more plentifully, and with more certainty than in any other; thus the *Shepherd's purse* has the greatest profusion of bloom in *April*, the *Daisy* in *May*, the *Poa annua* in *June*, and so on——Such we appre-hend to be the proper period of such plants flower-ing, and have set them down accordingly.

EXPLANA-

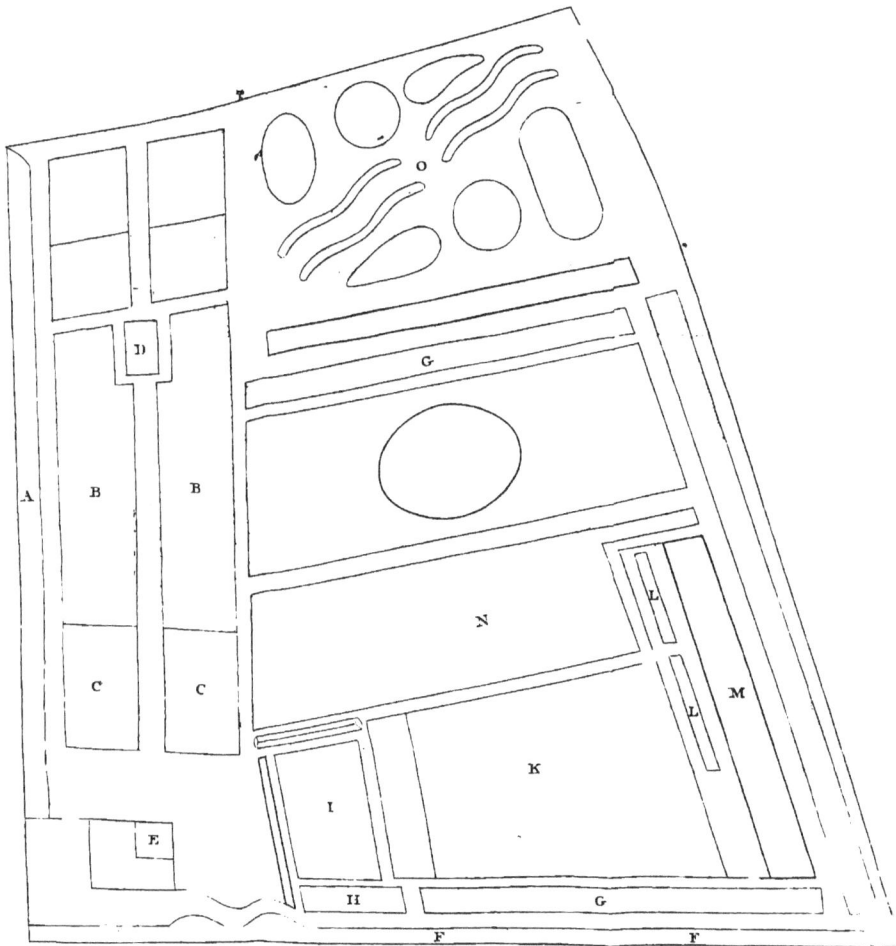

A _Tree & Shrub Quarter._ B _Agricultural Qu._ C _Class & Order Qu._ D _Greenhouse._ E _Library._ F _Culinary Qu._
G _Medicinal Qu._ H _Poisonous Qu._ I _Wood Qu._ K _Grass Qu._ L _Wall._ M _Water & Bog Qu._ N _General Q_
O _Ornamental Qu._

EXPLANATION of CHARACTERS, &c.

♄ Tree or Shrub.

♃ Perennial.

♂ Biennial.

☉ Annual.

L. Growing in the environs of London.

The following Abbreviations shew in what part of the Garden each Plant is to be found. Vid. *Plan of the Garden.*

Tr. Tree and Shrub Quarter.

G. General Quarter, containing the greatest assortment of British plants.

Gr. Grass Quarter.

B. Bog and Water Quarter.

W. Wall constructed with cavities, on a particular plan.

Wd. Wood Quarter, for plants growing in woods.

A

CATALOGUE

OF

BRITISH PLANTS,

Arranged according to their

PERIODS OF FLOWERING.

JANUARY.

Now Cæcias fends
His cutting winds, his ftorms of hail, and fnow,
Or binds the pregnant earth in icy chains;
The froft-nipt Laurel, and the blafted Bay,
Which lately fhone in all the pride of health,
Stand wither'd monuments of his dire reign.

. . .

FEBRUARY.

As yet the trembling year is unconfirm'd,
And winter oft, at eve refumes the breeze,
Chills the pale morn, and bids his driving fleets
Deform the day delightlefs.

Thomfon.

♄	1	Corylus Avellana	L. Hazel	Tr.
♃	2	Galanthus nivalis	Snowdrop	W.
♃	3	Helleborus viridis	Hellebore green	G.
♃	4	Tuffilago hybrida	Butterbur long-ftalked	G.

M A R C H.

Now fofter gales fucceed, at whofe kind touch,
Diffolving fnows in livid torrents loft,
The mountains lift their green heads to the fky.

Thomfon.

♄	5 Daphne Laureola	L. Spurge Laurel	Tr.
♄	6 ——— Mezereon	Mezereon	Tr.
♃	7 Helleborus fœtidus	Bear's foot	G.
♃	8 Juncus pilofus	L. Rufh fmall hairy Wood	B.
♃	9 Mercurialis perennis	L. Mercury Dog's	G.
♃	10 Rufcus aculeatus	L. Butcher's Broom	Wd.
♄	11 Taxus baccata	Yew	Tr.
♃	12 Viola odorata	L. Violet fweet	Wd.

A P R I L.

Call the vales, and bid them hither caft
Their bells, and flowrets of a thoufand hues.
Ye valleys low, where the mild whifpers ufe
Of fhades, and wanton winds, and gufhing brooks ;
On whofe frefh lap the fwart ftar fparely looks,
Throw hither all your quaint enamell'd eyes,
That on the green turf fuck the honied fhowers,
And purple all the ground with vernal flowers.

Milton.

☉	13 Arabis thaliana	L. Podded-Moufe-ear fmall	W
♃	14 Anemone nemorofa	L. Anemone white wood	Wd.
♃	15 ——— apennina	——— blue	Wd.
♂	16 Braffica Rapa	L. Turnep	G.
♂	17 ——— Napus	L. Navew wild	G.

♂	18 Cochlearia officinalis	Scurvy-grafs common	W.
	18 var. groenlandica	—— dwarf	W.
♂	19 —— danica	—— Danifh	W.
♂	20 —— anglica	L.—— fea	W.
☉	21 Cardamine hirfuta	L. Ladies-fmock hairy	W.
☉	22 Draba verna	L. Whitlow-grafs common	W.
♃	23 Equifetum fylvaticum	L. Horfe-tail wood	W.
♃	24 ———— arvenfe	L.———— corn	G.
♄	25 Fraxinus excelfior	L. Afh	Tr.
♃	26 Fragaria fterilis	L. Strawberry-barren.	G.
♃	27 Juncus campeftris	L. Rufh-hairy field	B.
♃	28 Narciffus Pfeudo-Narciffus	Daffodil common	W.
♃	Ornithogalum luteum	Star of Bethle'm yellow	W.
♄	29 Populus alba	Poplar white	Tr.
♄	30 —— tremula	L.—— trembling	Tr.
♄	31 —— nigra	L.—— black	Tr.
♄	32 —— Arbcel	L.—— Arbele	Tr.
♄	33 Prunus communis	L. Bullace Tree	Tr.
♄	34 —— fpinofa	L. Black thorn	Tr.
♄	35 —— Cerafus var. fructu nigro	L. Cherry wild —black	Tr.
♃	36 Primula vulgaris	L. Primrofe	Wd.
♃	37 Ranunculus Ficaria	L. Pilewort	G.
♃	38 ————auricomus	L. Crowfoot wood	G.
♄	*Salix hermaphroditica	Willow fhining	
♄	39 —— helix	L.—— fpurge-leaved	Tr.
	40 —— triandra	—— fmooth	Tr.
♄	41 —— Caprea	L.—— Sallow common	Tr.
♄	42 —— elongata	L.—— long-leaved	Tr.
♃	43 Scilla verna	Squil vernal	W.
♃	44 Saxifraga oppofiti-folia.	Saxifrage purple flowered	W.
♃	45 Tuffilago Farfara	L. Colt's foot	G.
♃	46 ———— Petafites	L. Butterbur common	G.
♃	47 Viola hirta	L. Violet hairy	Wd.
♃	Vaccinium uliginofum	Bilberry marfh	
♄	48 Ulmus campeftris var. folio latiffimo fcabro var. folio glabro	L. Elm common broad rough-leaved, or Wych Elm. fmooth-leaved	Tr.

Oft brush'd from Ruffia's wilds, a cutting gale
Rifes, and fcatters from his humid wings
The clammy mildew, or dry-blowing, breathes
Untimely froft, before whofe baleful blaft
The full-blown Spring, thro' all her foliage fhrinks
Joylefs and dead.

Thomfon.

M A Y.

Now the bright morning-ftar, day's harbinger,
Comes dancing from the Eaft, and leads with her
The flowery May, who from her green lap throws
The yellow cowflip, and the pale primrofe ;
   Hail bounteous May, that doft infpire
   Mirth, and youth, and warm defire ;
   Woods, and groves are of thy dreffing,
   Hill, and dale doth boaft thy bleffing:
Thus we falute thee with our early fong,
And welcome thee and wifh thee long.

Milton.

♃ 49	Anthoxanthum odo-ratum	L. Vernal grafs fweet-fcented	Gr.
♃ 50	Alopecurus pratenfis	L. Foxtail grafs Meadow	Gr.
♃ 51	Afperula odorata	L. Woodruff	W'd.
☉ 52	Afperugo procumbens	Madwort Goofe grafs	G.
☉ 53	Alfine media	L. Chickweed common	G.
♃ 54	Allium urfinum	Ramfons	W.
♃ 55	Adoxa Mofchatellina	L. Mofchatel tuberous	W.
♃ 56	Andromeda polifolia	L. Andromeda Rofemary-leaved	B.
♃ 57	Afarum europæum	Afarabacca	W'd.
♃ 58	Actæa fpicata	Bane berry	W'd.

♃ 59 Anemone Pulsatilla	Pasque-flower	W.
♃ 60 —— ranunculoides	Anemony yellow	Wd.
☉ 61 Arabis stricta	Podded-mouse-ear rough	W
♃ 62 Arum maculatum	L. Cuckow-pint	Wd.
♄ 63 Acer Pseudoplatanus	L. Maple Sycamore	Tr.
♄ 64 —— campestre	L. —— common	Tr.
♃ Anthericum serotinum	Anthericum bulbous	W.
♄ Arbutus alpina	Strawberry-tree alpine	Tr.
☉ 65 Bromus mollis	L. Brome-grafs foft	Gr.
☉ * —— muralis	—— wall	Gr.
♄ 66 Berberis vulgaris	L. Barberry	Tr.
♃ 67 Bellis perennis	L. Daify	G.
♄ 68 Buxus fempervirens	Box-tree	Tr.
♂ 69 Braffica oleracea	Cabbage fea	G.
♂ —— monenfis	—— Anglefea	G.
♃ 70 Chærophyllum fylveftre }	L. Cow-parfly common	G.
♂ 71 Carum Carui	Caraway	G.
♃ 72 Convallaria majalis	L. Lilly of the Valley	Wd.
♃ 73 —— Polygonatum	Solomon's-feal fweet	Wd.
♃ 74 —— multiflora	—— common	Wd.
♃ 75 Chryfofplenium op- pofitifolium	Golden-faxifrage L. common	B.
♃ 76 —— alternifolium	—— alternate-leaved	B.
☉ 77 Ceraftium vifcofum	L. { Moufe-ear-Chick- weed hoary }	W.
☉ 78 —— femidecandrum	L. —— fmall	W.
♄ 79 Cratægus Aria	Beam-tree white	Tr.
♄ 80 —— torminalis	L. Service-tree wild	Tr.
♄ 81 —— Oxyacantha	L. Hawthorn	Tr.
♃ 82 Caltha paluftris	L. Marfh-marigold	B.
♃ 83 Cochlearia Armoracia	L. Horfe-radifh	G.
♃ 84 Cheiranthus Cheiri	Wall-flower common	G.
♃ 85 Cardamine pratenfis	L. Ladies fmock common	W.
♃ 86 —— amara	L. —— bitter	W.
☉ 87 —— impatiens	—— impatient	W.
♃ —— petræa	—— mountain	W.
♃ * —— bellidifolia	—— daify-leaved	W.
♃ Carex faxatilis	Carex vernal	B.

♃	88 Carex panicea	L.	Carex pink	B.
♃	89 —— acuta	L.	—— acute	B.
♃	90 —— digitata		—— fingered	B.
♃	91 —— montana	L.	—— mountain	B.
♄	92 Carpinus Betulus	L.	Horn-beam Tree	Tr.
♃	93 Callitriche aquatica	L.	Starwort Water	B.
♃	94 Dentaria bulbifera		Coral-wort bulbiferous	Wd. Q.
☉	95 Draba muralis		Whitlow-grafs Speedwell-leaved	W.
♃	96 Empetrum nigrum		Crow-berry	B.
♃	97 Eriophorum vaginatum	L.	Cotton-grafs singleheaded	B.
♄	98 Euonymus europæus	L.	Spindle Tree	Tr.
♃	99 Euphorbia amygdaloides	L.	Spurge wood	Wd.
♂	100 Eryfimum Alliaria	L.	Saucealone	G.
♃	101 —— Barbarea	L.	Winter-crefs	G.
♃	102 Fragaria vefca	L.	Strawberry wood	G.
♃	103 Fritillaria Meleagris	L.	Fritillary common	W.
♄	104 Fagus fylvatica	L.	Beech	Tr.
♄	105* —— Caftanea	L.	Chefnut	Tr.
♃	106 Geum rivale		Avens Water	G.
♃	107 Glechoma hederacea	L.	Ground ivy	G.
☉	108 Geranium cicutarium	L.	Crane's-bill hemlock-leaved	G.
☉	—— pimpinellifolium		—— spotted flowered	G.
☉	109 —— mofchatum		—— mufk	G.
♃	110 —— phæum		—— dark flowered	
☉	111 —— molle	L.	—— common dove's foot	G.
☉	112 —— Robertianum	L.	Herb Robert	G.
♃	113 Hyacinthus nonfcriptus	L.	Hare-bell	W.
☉	Holoftea umbellata		Clove-chickweed	W.
♄	114 Hippophae rhamnoides		Sea Buckthorn	Tr.
♃	115 Juncus fylvaticus	L.	Rufh-great hairy Wood	B.

☉ 116 Iberis nudicaulis	L. Rock-Crefs	W.
♄ 117 Ilex Aquifolium	L. Holly	Tr.
118 Lamium album	L. Dead-Nettle white	G.
☉ 119 —— amplexicaule	L. —— Henbit	G.
☉ 120 —— purpureum	L. —— purple	G.
♃ 121 Leontodon Tarax- ⎱ acum ⎰	L. Dandelion common	G.
☉ 122 Lepidium petræum	Dittander rock	W.
♃ Lychnis Vifcaria	Lychnis Catchfly	G.
Lathræa Squamaria	Tooth-wort	Wd.
♃ 123 Menyanthes tri- ⎱ foliata ⎰	L. Buckbean	B.
♄ 124* Mefpilus germa- ⎱ nica ⎰	Medlar	Tr.
♄ 125 Myrica Gale	Gale fweet	Tr.
♃ 126* Narciffus poeticus	Daffodil pale	W.
♃ 127* Ornithogalum um- ⎱ bellatum ⎰	Star of Bethl'em com- mon	W.
♃ 128 Oxalis Acetofella	L. Wood-forrel	W.
♃ 129 Orobus tuberofus	L. Wood-pea	Wd.
♃ 130 Orchis mafcula	L. Orchis early-fpotted	W.
♃ 131 —— purpurea	—— purple man	W.
♃ —— militaris	—— man	W.
♃ 132 Ophrys aranifera	—— Spider	W.
♃ 133 Ophiogloffum ⎱ vulgatum ⎰	L. Adder's tongue	W.
♃ 134 Plantago lanceolata	L. Plantain Ribwort	G.
♃ 135 Pulmonaria offici- ⎱ nalis ⎰	Lungwort-fpotted	G.
♄ 136 Prunus Padus	L. Bird-Cherry	Tr.
♄ 137 Pyrus communis	L. Pear wild	Tr.
♄ 138 —— Malus	L. Crab	Tr.
♄ 149 Potentilla verna	Cinquefoil vernal	G.
♄ 140 Pinus fylveftris	Fir Scotch	Tr.
♄ 141* —— Picea	—— yew-leaved	Tr.
♄ 142* —— Abies	—— common	Tr.
♄ 143 Quercus Robur	L. Oak	Tr.
♄ 144 Rhamnus cathar- ⎱ ticus ⎰	L. Buckthorn	Tr.

		Latin		English	
♄	145	Rhamnus Frangula	L.	Berry-bearing Alder	Tr.
♄	146	Ribes rubrum		Currant red	Tr.
♄	147	—— nigrum		—— black	Tr.
♄	148	—— alpinum		—— alpine	Tr.
♃	149	Ranunculus bul-bofus	L.	Crowfoot bulbous	G.
♃	150	Rhodiola Rofea		Rofe root	G.
☉	151	Sherardia arvenfis	L.	Sherardia field	W.
☉	152	Sagina erecta	L.	Pearl-wort upright	W.
♃	153	Scandix odorata		Cicely fweet	G.
♂	154	Smyrnium Olu-fatrum	L.	Alexanders	G.
♄	155 *	Staphylea pinnata		Bladder-nut Tree	Tr.
☉	156	Saxifraga tridac-tylites	L.	Saxifrage rue-leaved	W.
♃	157	—— granulata	L.	—— white	W.
♃	158	Stellaria Holoftea	L.	Stichwort large	G.
♃	159	—— nemorum		—— wood	G.
♄	160	Sorbus aucuparia	L.	Mountain-Afh	Tr.
♄		—— domeftica		Service true	Tr.
♄	161	Schrophularia verna	L.	Figwort vernal	G.
♃	162	Sifymbrium Naf-turtium	L.	Water-Crefs	B.
☉	163	Senecio vulgaris	L.	Groundfel	G.
♄	164	Salix pentandra	L.	Willow fweet	Tr.
♄		—— herbacea		—— herbaceous	Tr.
♄	165	—— vitellina	L.	—— yellow-barked	Tr.
♄	166	—— amygdalina	L.	—— almond	Tr.
♄		—— reticulata		—— round-leaved	Tr.
♄	167	—— fragilis		—— crack	Tr.
♄		—— rubra		—— red	Tr.
♄	168	—— viminalis	L.	—— Ofier	Tr.
♄	169	—— repens	L.	—— creeping	Tr.
♄		—— arenaria		—— fand	Tr.
♄		—— rofmarinifolia		—— rofmary-leaved	Tr.
♄	170	—— alba	L.	—— white	Tr.
♄		—— lanata		—— woolly	Tr.
♄		—— aurita		—— round-eared	Tr.

M

☉	171	Thlaspi Bursa pastoris	L.	Shepherd's Purse	G.
☉		Tillæa muscosa		Tillæa mossy	W.
♄	172	Ulex europæus	L.	Furze, Whins, Gorse,	Tr.
☉	173	Veronica arvensis	L.	Speedwell field	W.
☉	174	——— hederifolia	L.	——— ivy-leaved	G.
♃	175	Valeriana dioica	L.	Valerian Marsh	G.
☉	176	——— locusta	L.	Corn-sallad	G.
♃	177	Vinca minor	L.	Periwinkle small	G.
♄	178	Viburnum Lantana	L.	Wayfaring Tree	Tr.
♄	179	——— Opulus	L.	Guelder Rose	Tr.
♃	180	Valantia Cruciata		Crofswort	G.
♃	181	Vaccinium Myrtillus	L.	Whortleberry common.	B.
♃	182	——— Vitis idæa		Bilberry red	B.
♃	183	——— Oxycoccos		Cranberry	B.
♃	184	Viola canina	L.	Violet dogs	Wd.
♃	185	——— paluftris	L.	——— bog	B.
☉		Veronica triphyllos		Speedwell trifid	W.
☉		——— verna		——— vernal	W.
♃		Viscum album		Misletoe	Wd.

J U N E.

No gradual bloom is wanting, from the bud,
First born of spring, to summer's musky tribes,
Nor Hyacinths of purest virgin white,
Low-bent and blushing inward, nor Jonquils
Of potent fragrance, nor Narcissus fair,
As o'er the fabled fountain hanging still,
Nor broad Carnations, nor gay spotted Pinks;
Nor showr'd from every bush, the damask Rose,
Infinite numbers, delicacies, smells,
With hues on hues, expreffion cannot paint
The breath of Nature, and her endless bloom.
                          Thomson.

☉	186 Æthufa Cynapium	L. Fool's Parfly	G.
♃	187 —— Meum	Spignel	G.
♃	188 Ægopodium Po- dagraria }	L. Gout-weed	G.
♃	Afperula Cynanchica	Squinancy-wort	W.
♃	189 Alchemilla vulgaris	Ladies-mantle common	G.
☉	190 Aphanes arvenfis	L. Parfly piert	W.
☉	191 Anchufa femper- virens }	Alkanet ever-green	G.
♃	192 Antirrhinum Cym- balaria	L. Toad-flax ivy-leav- ed	W.
☉	193 —— Elatine	L. { Fluellin fharp-point- ed	W.
♃	194 —— majus	L. Snapdragon large	W.
♂	195 Arabis turrita	{ Podded moufe-ear tower	W.
☉	196 Arenaria trinervia -	L. { Chickweed plantain- leaved	W.
☉	197 —— ferpyllifolia	L. —— thyme-leaved	W.
♃	198 —— verna	—— mountain	W.
☉	199 —— tenuifolia	—— fine-leaved	W.
☉	200 —— rubra	L. —— red-flowered	W.
☉	201 —— marina	—— fea	W.
☉	202 Agroftema Githago	L. Cockle	G.
♃	203 Afparagus officinalis	Afparagus	G.
♃	204 Acorus Calamus	L. Sweet-Cane	B.
♃	205 Aquilegia vulgaris	L. Columbine's common	G.
♂	206 Adonis autumnalis	Pheafant's eye	G.
♃	207 Ajuga reptans	L. Bugle	G.
♃	208 Aftragalus arenarius	Milk-vetch purple	W.
♃	209 Avena pratenfis	Oat-grafs meadow	Gr.
♃	210 —— pubefcens	L. —— rough	Gr.
♃	211 —— flavefcens	L. —— yellow	Gr.
♃	212 —— elatior	L. —— tall	Gr.
☉	213 Ægilops incurva	Hard-grafs fea	Gr.
☉	214 Alopecurus agreftis,	L. Foxtail-grafs field	Gr.
♃	215 Aira aquatica	L. Hair-grafs water	Gr.
☉	216 —— præcox	L. —— early	Gr.
☉	217 —— caryophyllea	L. —— filver	Gr.

♃	218 Allium arenarium	Garlic fand	W.
♃	219——— Schænoprafum	Cives	W.
♄	Arbutus Uva Urfi	Bear-berry	Tr.
♃	*Aquilegia alpina	Columbine mountain	G.
♃	Ajuga pyramidalis	Bugle mountain	G.
♃	Azalea procumbens	Azalea procumbent	G.
♃	220 Arundo arenaria	Reed-grafs fea	Gr.
☉	*Bufonia tenuifolia	Bufonia fine-leaved	W.
♃	221 Bromus erectus	Brome grafs upright	Gr.
☉	222 ——— fterilis	L. ——— barren	Gr.
☉	Bunias Cakile	Rocket fea	G.
♂	223 Borago officinalis	L. Borage	G.
♃	224 Bunium Bulbo-caftanum }	L. Earth or Pig-nut	Wd.
♄	225 Betula alba	L. Birch common	Tr.
♄	226 ——— nana	——— dwarf	Tr.
♄	226 ——— Alnus	L. ——— Alder	Tr.
♄	227 Bryonia alba	L. Bryony white	G.
♃	228 Carex *depauperata*	L. Carex *Charlton*	B.
♃	229 ——— diftans	L. ——— loofe	B.
♃	230 ——— fpicata	L. ——— fpiked	B.
♃	231 ——— inflata	——— bottle	B.
♃	232 ——— cæfpitofa	L. ——— turfy	B.
♃	233 ——— veficaria	L. ——— bladder	B.
♃	234 ——— hirta	L. ——— hairy	B.
♃	235 ——— pulicaris	L. ——— flea	B.
♃	236 —— paniculata	L ——— panicled	B.
♃	237 ——— diftichla	.. ——— foft	B.
♃	238 ——— arenaria	——— fea	B.
♃	239 ——— leporina	L. ——— naked	B.
♃	240 ——— vulpina	.. ——— great	B.
♃	——— divifa	——— marfh	B.
♃	241 ——— canefcens	——— grey	B.
♃	242 ——— muricata	L. ——— prickly	B.
♃	243 ——— remota	L. ——— remote	B.
♃	244 ——— flava	L. ——— yellow	B.
♃	245 ——— bulifera	L. ——— globular	B.
♃	246 ——— pallefcens	L. ——— pale	B.
♃	247 ——— pendula	L. ——— pendulous	B.

♃	248	Carex fylvatica	L.	Carex wood	B.
♃		—— dioica		—— fmall	B.
♃		—— capitata		—— round-headed	B.
♃	249	—— montana	L.	—— mountain	B.
♃		—— atrata		—— black	B.
♃		—— paucifloa		—— few-flowered	B.
♃		—— brizoides		—— rough	B.
♃		—— limofa		—— brown	B.
♃		—— ftrigofa		—— loofe	B.
♃		—— recurva		—— héath	B.
♃	250	—— *riparia*	L.	—— *common*	B.
♃	251	—— *gracilis*	L.	—— *flender fpiked*	B.
♃	252	—— *juncea*		—— *rufh*	B.
♃	253	—— capillaris		—— capillary	B.
♄	254	Cornus fanguinea	L.	Dogweed common	Tr.
♄		—— herbacea		—— herbaceous	Tr.
☉		Caucalis daucoides		Caucalis rough	G.
☉	255	—— anthrifcus	L.	—— hedge	G.
☉	256	—— arvenfis	L.	—— field	G.
☉	257	—— nodofa	L.	—— knotted	G.
♃		Cucubalus acaulis		Campion mofs	W.
♃	258	Cotyledon Umbilicus		Navel-wort common	W.
♃	*	—— lutea		—— yellow	W.
♃		Ceraftium alpinum		{ Moufe-ear-chickweed alpine	W.
♃	*	—— latifolium		—— broad-leaved	W.
♃	259	—— vulgatum	L.	—— common	W.
♃	260	—— arvenfe	L.	—— creeping	W.
☉		Ciftus guttatus		Ciftus annual	W.
♃	261	—— hirfutus		—— hairy	W.
♃		—— falicifolius		—— willow-leaved	W.
♂		Cheiranthus finuatus		Stock fea	G.
♂	262	—— eryfimoides		—— yellow	G.
♂		Cineraria paluftris		Cineraria marfh	B.
♃	263	—— alpina		—— mountain	W.
♃	264	Cynofurus cæruleus		Dog's tail-grafs blue	Gr.
♃	265	Campanula glomerata		Bell-flower cluftered	G.

☉ 266 Campanula hybrida	L. Bell-flower corn	G.
♃ 267 Chærophyllum temulum	L. Cow-parfly fmall	G.
♃ 268 Comarum paluftre	Marfh-Cinquefoil	B.
☉ 269 Cochlearia Coronopus	L. Swine's-Crefs common	W.
☉ 270 ——— didyma	——— double-podded	W.
♃ 271 Crambe maritima	Colewort fea	G.
♃ 272 Cypripedium Calceolus	Ladies flipper	W.
♃ 273 Dryas octopetala	Dryas mountain	W.
☉ 274 Delphinium Confolida	Larkfpur	G.
♃ 275 Doronicum Pardalianches	Leopard's bane	G.
♃ 276 Eriophorum polyftachion	L. Cotton-grafs many headed	B.
☉ 277 Erigeron acre	L. Erigeron biting	G.
☉ 278 Ervum tetrafpermum	L. Tine-tare fmooth podded	G.
☉ 279 ——— hirfutum	L. ——— hairy-podded	G.
♃ 280 Equifetum limofum	L. Horfetail fmooth	B.
♃ 281 ——— paluftre	——— marfh	B.
♃ 282 ——— fluviatile	L. ——— river	B.
♃ 283 Feftuca duriufcula	L. Fefcue-grafs hard	Gr.
☉ 284 ——— bromoides	L. ——— barren	Gr.
☉ ——— uniglumis	L. ——— fingle-glumed	Gr.
☉ 285 ——— myuros	L. ——— wall	Gr.
♃ 286 ——— fluitans	L. ——— flote	Gr.
☉ 287 Fumaria officinalis	L. Fumitory common	G.
☉ Galium fpurium	L. Cleaver fmooth-feeded	G.
♃ 288 ——— Aparine	L. ——— common	G.
☉ 289 ——— montanum	L. Bedftraw mountain	W.
☉ Galeopfis Ladanum	Nettle-hemp red	G.
☉ 290 ——— Tetrahit	L. ——— common	G.
♃ 291 Galeobdolon Galeopfis	L. Dead-nettle yellow	G.
♃ 292 Gnaphalium dioicum	Cudweed dioicous	W.

♃	293 Glaux maritima	Sea-milkwort	B.
♃	294 Geum urbanum	L. Avens common	G.
☉	295 Geranium mariti-mum	Crane's-bill fea	W.
♃	296 ——— nodofum	——— knotted	G.
♃	297 ——— fylvaticum	——— wood	G.
☉	298 ——— lucidum	L. ——— fhining	G.
☉	299 ——— columbinum	L. ——— long-ftalked	G.
☉	300 ——— diffectum	L. ——— jagged	G.
♂	Hefperis inodora	Dame s-Violet	G.
♃	301 Hippuris vulgaris	L. Mare's-tail	B.
♃	302 Hordeum murinum	L. Barley-grafs wall	Gr.
♃	303 Hydrocotyle vulgaris	L. Pennywort	B.
♃	304 Hippocrepis comofa	Horfe-fhoe Vetch	W.
☉	305 Hypochæris glabra	L. Hawkweed fmall-flowered	G.
♃	——— maculata	——— fpotted	G.
♃	306 ——— radicata	L. ——— long-rooted	G.
♃	307 Humulus Lupulus	L. Hop	G.
♃	308 Hydrocharis Mor-fus ranæ	L. Frog-bit	B.
♃	309 Imperatoria Of-truthium	Mafter-wort	G.
♃	Juncus trifidus	Rufh trifid	B.
♃	——— triglumis	——— three-glumed	B.
♄	310 Juniperus communis	L. Juniper common	Tr.
♃	311 Ifatis tinctoria	Woad	G.
☉	312 Jafione montana	L. Sheep's-fcabious hairy	W.
♄	313 Liguftrum vulgare	L. Privet	Tr.
♃	314 Lolium perenne	L. Rye or Ray-grafs	Gr.
♃	316 Lyfimachia ne-morum	L. Pimpernel yellow	Wd.
☉	316 Lycopfis arvenfis	L. Buglofs fmall	G.
♃	317 Lithofpermum pur puro-cæruleum	Gromwell purple	G.
☉	319 ——— arvenfe	L. ——— corn	G.
♃	320 Lychnis Flos Cu-culi	L. Ragged Robin	G.

♃	320	Lychnis dioica	L. Lychnis red	G.
♃	321	——— *alba*	L. ——— *white*	G.
⊙	322	Lepidium ruderale	Dittander narrow-leaved	G.
⊙	323	Lathyrus Niffolia	L. Grafs-vetch crimfon	G.
⊙	324	——— Aphaca	L. Vetchling yellow	G.
♃	325	Littorella lacuftris	L. Littorella marfh	B.
♃	326	Melica nutans	L. Melic-grafs common	Gr.
♃	327	——— montana	——— mountain	Gr.
⊙	328	Montia fontana	L. Blinks	B.
⊙	329	Myofotis fcorpioides	L. Moufe-ear fcorpion grafs	G
		var. fl. luteis	L. ——— yellow-flowered	G.
♃	330	——— *paluftris*	L. ——— *marfh*	B.
⊙	331	Myofurus minimus	L. Moufe-tail	W.
♃	332	Melittis Melif- fophyllum	Baftard-Balm	G.
⊙	333	*Myagrum fativum	Gold of pleafure	G.
♃	334	Medicago lupulina	L. Medick hop, Nonefuch	G.
♃		Orobanche major	L. Broom-rape common	
♃		——— ramofa	——— branched	
♃		*Orchis coriophora	Orchis fmall lizard	W.
♃	335	——— Morio	L. ——— meadow	W.
♃	336	——— uftulata	L. ——— dwarf	W.
♃	337	——— latifolia	——— marfh	W.
♃		Ophrys Nidus avis	——— bird's-neft	W.
♃		——— cordata	——— heart-fhaped	W.
♃		——— paludofa	——— bog	W.
♃		——— lilifolia	——— lily-leaved	W.
♃	338	——— mufcifera	——— fly	W.
♃	339	——— apifera	L. ——— bee	W.
♃	340	——— anthropophora	L. ——— Tway man	W.
♃	341	——— ovata	——— blade	W.
⊙	342	Ornithopus per- pufillus	L. Bird's foot true	W.
♃	343	Ornithogalum py- reniacum	Star of Bethl'em moun- tain	W.
♃	344	Ononis fpinofa	L. Reftharrow prickly	G.
♃	345	——— arvenfis	L. ——— fmooth	G.
♃		Orobus fylvaticus	Vetch wood	Wd.
♃		Ofmunda Lunaria	L. Moonwort	W.

♃	346	Oenanthe pimpi-nelloides	L.	Dropwort Meadow	B.
♃	347	Oxalis corniculata		Wood-Sorrel yellow	W.
♃		Pinguicula villosa		Butterwort small	B.
♃	348	—— vulgaris		—— common	B.
♃		*Poa bulbosa	L.	Meadow-grafs bulbous	Gr.
♃	349	—— pratensis	L.	—— fmooth-ftalked	Gr.
♃	350	—— trivialis	L.	—— rough-ftalked	Gr.
♃		—— anguftifolia	L.	—— narrow-leaved	Gr.
♃	351	—— nemoralis	L.	—— wood	Gr.
☉	352	—— annua	L.	—— annual	Gr.
☉	353	—— rigida	L.	—— hard	Gr.
♃	354	—— *retroflexa*		—— *reflexed*	Gr.
♃		Potamogeton fer-ratum	L.	Pondweed ferrated	B.
♃		Pulmonaria mari-tima		Lungwort fea -	G.
♃		Pimpinella dioica		Burnet-faxifrage leaft	W.
♃	355	Pyrola minor		Winter-green fmall	W.
♃		—— fecunda		—— ferrated	W.
☉	356	Papaver hybridum	L.	Poppy round prickly-headed	G.
☉	357	—— Argemone	L.	—— long prickly-headed	G.
☉	358	—— Rhœas	L.	—— round fmooth-headed	
☉	359	—— dubium	L.	—— long fmooth-headed	G.
♃	360	—— cambricum		—— yellow	G.
☉	361	—— fomniferum	L.	—— garden	G.
♃	362	Peucedanum offi-cinale		Hog's-fennel	G.
♃	363	Primula veris	L.	Cowflip	Wd.
♃	364	—— *inodora*	L.	*Oxlip* -	Wd.
♃	365	—— farinofa		Bird's-eye	B.
♃	366	Polygonum Biftorta	L.	Biftort common	G.
♃	367	—— viviparum		—— fmall	G.
♃	368	Paris quadrifolia	L.	Herb Paris	Wd.
♄	369	Potentilla fruticofa		Cinquefoil fhrubby	Tr.

N

♃	370 Potentilla Anferina	L. Silver weed	G.
♃	371 —— rupeftris	—— rock	G.
♃	372 —— argentea	L. —— tormentil	G.
♃	373 —— alba	—— white	G.
♃	374 —— reptans	L. —— common	G.
♃	375 Pedicularis fylvatica	L. Red-rattle heath	B.
♃	Pilularia globulifera	L. Pepper-grafs	B.
♃	376 Refeda lutea	Bafe-rocket	G.
☉	377 —— Luteola	L. Dyer's-weed	G.
♃	Rubus Chamæmorus	Cloud-berry	
♃	378 —— faxatilis	Bramble ftone	G.
♃	379 Rhinanthus Crifta galli	L. { Yellow-rattle	B.
♃	380 Rumex Acetofa	L. Sorrel common	G.
♃	381 —— Acetofella	L. —— fheep's	G.
♄	382 Rofa rubiginofa	L. Rofe fweet-briar	Tr.
♄	383 —— fpinofiffima	L. —— burnet	Tr.
♄	384 —— arvenfis	L. —— white	Tr.
♄	385 —— villofa	L. —— apple	Tr.
♄	386 —— canina	L. —— dog	Tr.
♃	387 Ranunculus acris	L. Crowfoot tall	G.
☉	388 —— fceleratus	L. —— celery-leaved	G.
☉	389 —— parviflorus	L. —— fmall-flowered	G.
☉	390 —— arvenfis	L. —— corn	G.
☉	391 —— hederaceus	L. —— ivy-leaved	B.
♃	392 —— aquatilis	L. —— water	B.
☉	393 Raphanus Raphaniftrum }	L. Radifh wild	G.
☉	Silene noctiflora	{ Catchfly night-flowering	G.
☉	—— conoidea	—— corn	G.
☉	394 —— quinquevulnera	—— variegated	G.
♃	395 —— nutans	—— Nottingham	G.
♃	—— amæna	—— fea	G.
☉	396 —— Armeria	—— common	G.
♄	Sorbus hybrida	Service baftard	Tr.
♃	Satyrium hircinum	Satyrion great lizard	W.
♃	—— albidum	—— white	W.
♃	—— repens	—— creeping	W.

♃ 397* Salvia pratensis	Clary meadow	G.
☉ 398 —— verbenaca	L. —— wild	G.
♃ 399 Sagina procumbens	L. Pearlwort procumbent	W.
☉ 400 —— *apetala*	L. —— *apetalous*	W.
♃ 401 Symphytum officinale	L. Comfrey common	G.
♃ 402 ———tuberosum	—— tuberous	G.
♃ 403 Sanicula europæa	L. Sanicle	Wd.
☉ 404 Scandix Pecten	L. Shepherd's needle	G.
☉ 405 —— Anthriscus	L. Scandix rough-seeded	G.
♄ 406 Sambucus nigra	L. Elder	Tr.
♃ 407 Sibbaldia procumbens	Sibbaldia procumbent	W.
☉ 408 Scleranthus annuus	L. Knawel annual	W.
☉ —— perennis	—— perennial	W.
♃ 409 Saxifraga stellaris	Saxifrage hairy	B.
♃ 410 —— hirculus	—— marsh	B.
♃ 411 —— hypnoides	—— trifid	W.
♃ 412 Stellaria graminea	L. Stichwort small	G.
♃ 413 Sisymbrium sylvestre	L. Rocket creeping	B.
♃ 414 —— amphibium	L. Water-radish amphibious	B.
415 —— *terrestre*	L. —— *annual*	B.
☉ 416 —— Irio	L. Rocket London	G.
☉ 417 —— Sophia	L. Flixweed	G.
☉ 418 Sinapis arvensis	L. Charlock	G.
☉ 419 Spartium scoparium	L. Broom	Tr.
♄ 420 Sonchus oleraceus	L. Sow-thistle	G.
☉ 421 *Senecio viscosus*	*Groundsel stinking*	G.
☉ 422 —— sylvaticus	L. —— mountain	G.
☉ Salix Myrsinites	Willow whortle-leaved	Tr.
♄ —— lapponum	—— wooly	Tr.
♄ Thesium linophyllum	Bastard-Toad-flax	W.
☉ *Tordylium officinale	Tordylium officinal	G.
♃ Trientalis europæa	Chickweed-Wintergreen	W.

♂	Thlaſpi alpeſtre	{	Mithridate-muſtard alpine	W.
☉ 423	—— campeſtre	——	common	G.
☉ 424	—— arvenſe	——	broad-podded	G.
♃ 425	—— montanum	——	mountain	W.
☉	Trifolium Melilotus	L.	Melilot	G.
♃ 426	—— ſcabrum		Trefoil rough	W.
☉ 427	— ornithopodioides	L.	—— bird's foot	W.
♃ 428	—— pratenſe	L.	—— meadow	G.
☉ 429	—— ſubterraneum	L.	—— ſubterraneous	W.
♂ 430	Tragopogon pra-tenſe }	L.	Goat's-beard	G.
♂ 431	—— porrifolium	L.	Salſafy	G.
♂ 432	Triticum unila-terale }	L.	Wheat-graſs dwarf	Gr.
♂ 433	Turritis glabra	L.	Tower-muſtard ſmooth	G.
♃ 434	—— hirſuta		—— hairy	G.
♃ 435	Trollius europæus		Globe-flower	G.
♃ 436	Tamus communis	L.	Black-Bryony	G.
♃ 437	Tormentilla offi-cinalis }	L.	Tormentil	G.
438	Veronica montana	L.	Speedwell mountain	G.
439	—— ſerpyllifolia	L.	—— thyme-leaved	G.
440	—— Chamædrys	L.	—— Germander-leaved	G.
441	—— agreſtis	L.	—— Chickweed	W.
442	Valeriana rubra		Valerian red	G.
443	Vinca major	L.	Periwinkle large	G.
444	Vella annua		Creſſe-rocket	W.
445	Vicia ſativa	L.	Vetch, or Tare common	G.
446	—— lathyroides		—— ſpring	G.
447	Viola tricolor	L.	Panſie common	Wd.
448	—— grandiflora		—— yellow	Wd.

J U L Y.

Deep to the root
Of vegetation parch'd, the cleaving fields
And flipp'ry lawn an arid hue diſcloſe ;
Echo no more returns the cheerful found
Of ſharpning ſcythe ; the mower finking heaps
O'er him the humid hay, with flowers perfum'd.

Thomſon.

24 449	Agroſtis littoralis	Bent-graſs Norfolk	Gr.
24	—— ſetacea	—— ſheep's-feſcue	Gr.
24	—— *pumila*	—— *dwarf*	Gr.
⊙ 450	—— ſpica venti	L. —— bearded	Gr.
24 451	—— canina	L. —— brown	Gr.
24 452	—— *tenuifolia*	L. —— *fine-leaved*	Gr.
24 453	—— *alba*	L. —— *white*	Gr.
24 454	—— capillaris	L. —— fine-panicled	Gr.
24 455	—— *ſtolonifera*	L. —— *couchy*	Gr.
24	Arundo epigejos	Reed-graſs ſmall	Gr.
24 456	Aſperula cynanchica	Squinancy wort	W.
24	Athamanta Oreoſelinum	Spignel mountain	G.
24 457	Allium Ampeloprafum	Garlic round-headed	G.
	—— oleraceum	—— herbaceous	G.
	Aliſma natans	Water-Plantain creeping	B.
458	—— Plantago aquatica	L. —— great	B.
459	—— ranunculoides	—— ſmall	B.
460	—— Damaſonium	L. —— ſtarry-headed	B.
	Arenaria laricifolia	Chickweed larch-leaved	W.
461	—— peploides	—— purſlane-leaved	W.
462	Antirrhinum repens	Toad-flax creeping	G.
463	—— arvenſe	—— corn	G.

464 Antirrhinum Linaria	L. Toad-flax common	G.
465 —— minus	—— fmall	
Aftragalus uralenfis	Aftragalus hairy	
466 —— glycyphyllos	L. Liquorice Vetch	G.
Artemifia campeftris	Southernwood field	G.
Anthemis arvenfis	Mayweed corn	G.
—— maritima	—— fea	G.
467 —— Cotula	L. —— ftinking	G.
468 —— tinctoria	Ox-eye	G.
469 Alopecurus geniculatus	L. Fox-tail jointed	Gr.
470 —— bulbofus	—— bulbous	Gr.
471 —— monfpelienfis	—— bearded	Gr.
472 Aira cefpitofa	L. Hair-grafs turfy	Gr.
473 —— flexuofa	L. —— heath	Gr.
474 —— fetacea	—— fine-leaved	Gr.
475 —— canefcens	—— grey	Gr.
476 Avena nuda	Oat-grafs naked	Gr.
477 —— fatua	—— bearded	Gr.
478 Alchemilla alpina	Ladies-mantle alpine	W.
479 Anagallis arvenfis	L. Pimpernel common	G.
480 —— tenella	L. —— bog	B.
481 Atropa Belladonna	Deadly Nightfhade	G.
482 Anethum Fœniculum	Fennel	G.
483 Apium graveolens	L. Smallage	G.
484 Althæa officinalis	Marfh-mallow	G.
485 Anthyllis vulneraria	Ladies-finger	W.
486 Arctium Lappa	L. Burdock	G.
487 Achillea Millefolium	L. Yarrow	G.
488 Ariftolochia Clematitis	Birthwort climbing	G.
489 Anthericum offifragum	L. Afphodel Lancafhire	B.
♃ 490 Briza media	L. Quaking-grafs common	Gr.
☉ 491 —— minor	—— fmall	Gr.
☉ 492 Bromus fecalinus	L. Brome-grafs corn	Gr.

♃	493	Bromus giganteus	L. Brome-grafs tall	Gr.
♃	494	—— hirfutus	L. —— hairy-ftalked	Gr.
♃	495	Beta maritima	Beet fea	G.
☉	496	Bupleurum rotun-difolium	L. Thorough-wax common	G.
☉	497	—— tenuiffimum	L. —— leaft	G.
♃	498	Butomus umbellatus	L. Flowering-rufh	B.
♃	499	Betonica officinalis	L. Betony wood	G.
♃	500	Braffica muralis	L. Rocket ftinking	G.
☉		—— campeftris	Cabbage field	G.
☉		—— orientalis	—— perfoliate	G.
♂	501	Campanula patula	Bell-flower fpreading	G.
♂	502	—— Rapunculus	L. —— Rampion	G.
♃	503	—— latifolia	Throat-wort Giant	G.
♃	504	—— Trachelium	L. —— common	G.
♃	505	—— hederacea	L. Bell-flower ivy-leaved	W.
☉		Cufcuta europæa	Dodder	
☉		Caucalis latifolia	Caucalis broad-leaved	G.
♂		Chelidonium hybridum }	L. Horn-poppy violet	G.
♂		—— corniculatum	—— red	G.
♂	506	—— Glaucium	L. —— yellow	G.
♃	507	—— majus	L. Celandine common	G.
♃		Crepis fœtida	{ Succory-Hawkweed ftinking	G.
♂	508	—— tectorum	L. —— common	G.
♂	509	—— biennis	L. —— Kentifh	G.
♃	510	Circæa lutetiana	L. { Enchanter's Night-fhade common	G.
♃	511	—— alpina	—— alpine	G.
♃	512	Cynofurus criftatus	L. Dog's-tail-grafs crefted	Gr.
☉	513	Centunculus minimus }	L. Baftard-Pimpernel	B.
♂	514	Cynogloffum officinale }	L. Hound's-tongue	G.
♃	515	Convolvulus arvenfis	L. Bindweed fmall	G.
♃	516	—— Soldanella	—— fea	G.
☉	517	Chironia Centaurium }	L. Centaury	W.

☉	518 Chenopodium Vulvaria }	L. Orach ſtinking	G.
☉	519 —— polyſpermum	L. Allſeed	G.
♂	520 Conium maculatum	L. Hemlock	G.
♃	521 Crithmum maritimum }	Samphire	G.
♃	522 Cicuta viroſa	L. { Water-Hemlock long-leaved	B.
♂	523 Coriandrum ſativum	Coriander	G.
♂	524 Chlora perfoliata	Yellow-wort perfoliate	W.
♃	525 Cucubalus bacciferus	Campion berry-bearing	G.
♃	526 —— Behen	L. —— bladder	G.
♃	—— viſcoſus	—— Dover	W.
♃	527 —— Otites	—— Spaniſh	W.
♃	Cherleria ſedoides	Cherleria moſſy	W.
♃	528 Ciſtus Helianthemum }	L. Ciſtus dwarf	W.
♃	529* —— ſurrejanus	—— narrow-petal'd	W.
♃	530 —— polifolius	—— mountain	W.
♃	531 Clinopodium vulgare }	L. Baſil wild	G.
♃	532 Cichorium Intybus	L. Succory blue	G.
♃	533 Carduus arvenſis	L. Thiſtle curſed	G.
♂	534 —— lanceolatus	L. —— ſpear	G.
♂	535 —— nutans	L. —— muſk	G.
♂	536 —— criſpus	L. —— curled	G.
☉	537 —— acanthoides	L. —— welted	G.
♂	538 —— paluſtris	L. —— marſh	G.
♃	539 —— heleniodes	—— melancholy	G.
♃	540 —— heterophyllus	L. —— meadow	G.
♂	541 —— marianus	L. —— milk	G.
♂	—— eriophorus	—— wooly-headed	G.
♃	542 —— acaulis	—— dwarf	G.
♂	543 Carlina vulgaris	L. —— Carline	G.
☉	544 Chryſanthemum ſegetum }	L. Corn-marigold	G.
♃	545 —— Leucanthemum	L. Ox-eye-daiſy	G.
☉	546 Centaurea Cyanus	L. Blue-bottle	G.

♂	547	Centaurea sol-stitialis		St. Barnaby's Thistle	G.
♃		Carex tomentosa		Carex downy	B.
♃	548	— pseudo-Cyperus	L.	— baftard Cyperus	B.
☉		Chara vulgaris	L.	Chara common	
☉		—— tomentosa		—— downy	
☉		—— hispida		—— hispid	
☉		—— flexilis	L.	—— flexible	
☉		Drosera rotundi-folia	L.	Sun dew round-leaved	B.
☉		—— longifolia	L.	—— long-leaved	B.
♃		Dianthus arenaria		Pink common	G.
☉	549	—— Armeria	L.	—— Deptford	G.
☉	550	—— prolifer		—— proliferous	G.
♃	551	—— Caryophyllus		—— Clove	G.
♃	552	—— deltoides	L.	—— maiden	W.
♃	553	—— glaucus		—— mountain	W.
♃	554	Dactylis glomeratus	L.	Cock's-foot-grafs rough	Gr.
♂	555	Daucus Carota	L.	Wild-Carrot	G.
♂	556	Draba incana		Whitlow-grafs wreathen-podded	W.
♂	557	Digitalis purpurea	L.	Fox-glove	G.
♂		Echium italicum		Viper's-Buglofs wall	G.
♂	558	—— vulgare	L.	—— common	G.
♃		* Echinophora spinosa		Prickly-famphire	G.
♂		Euphorbia Paralias		Spurge fea	W.
♃	559	Elymus arenarius		Lyme-grafs fea	G.
♃	560	—— caninus	L.	—— dogs	Gr.
♃	561	Eryngium mari-timum		Eryngo fea	G.
♃	562	—— campeftre		—— common	G.
♃	563	Epilobium anguf-tifolium		Willow-herb Rofebay	G.
♃	564	—— hirfutum	L.	—— large-flowered	G.
♃	565	—— villofum	L.	—— hoary	G.
♃	566	—— montanum	L.	—— mountain	G.
♃	567	—— *dubium*	L.	—— *fpurious*	G.
♃	568	—— tetragonum	L.	—— fquare ftalked	G.

O

		Latin		English	
♃	569	Epilobium paluſtre	L.	Willow-herb marſh	B.
♃	570	—— alpinum		—————— alpine	W.
♃	571	Erica cinerea	L.	Heath fine-leaved	B.
♃	572	—— tetralix	L.	—— croſs-leaved	B.
♂	573	Eryſimum officinale	L.	Hedge-muſtard	G.
♂	574	—— cheiranthoides		Treacle-wormſeed	G.
♃	575	Eupatorium canna- binum	L.	Water-Hemp-agri- mony	G.
♃		Eriocaulon de- cangulare		Networt	B.
♃		Frankenia pul- verulenta		Sea-Heath duſty	W.
♃	576	—————— lævis		—————— ſmooth	W.
♃	577	Feſtuca ovina	L.	Feſcue-graſs ſheep's	Gr.
♃	578	—— var. vivipara		—— viviparous	Gr.
♃	579	—— prateuſis	L.	—— meadow	Gr.
♃	580	—— elatior	L.	—— tall	Gr.
♃	581	—— pinnata	L.	—— ſpiked	Gr.
♃	582	—— decumbens	L.	—— decumbent	Gr.
♃	583	—— ſylvatica	L.	—— wood	Gr.
♃		—— glabra	L.	—— ſmooth	Gr.
♂	584	Fumaria claviculata	L.	Fumitory	Gr.
♃		Galium erectum		Ladies-bedſtraw upright	B.
♃	585	—— paluſtre	L.	—— marſh	B.
♃	586	—— uliginoſum	L.	—— bog	B.
♃	587	—— Mollugo	L.	—— white	G.
♃	588	—— verum	L.	—— yellow	G.
♃		Galeopſis villoſa		Nettle-hemp hairy	G.
☉		Gnaphalium lu- teo-album		Cudweed Jerſey	W.
♃	589	Gentiana Pneu- monanthe	L.	Calathian Violet	W.
♃	590	Geranium pratenſe	L.	Crane's-bill Crowfoot	G.
♃	591	—— pyrenaicum	L.	—— mountain	G.
☉	592	—— rotundifolium	L.	—— true Dove's-foot	G.
♃	593	—— ſanguineum	.	—— bloody	G.
♃	594	Geniſta tinctoria	L.	Greenweed Dyers	G.
♃	595	—— piloſa		—— hairy	W.
♄	596	—— anglica	L.	—— Petty Whin	Tr.

♃	Hieracium alpinum	Hawkweed alpine	W.
♃	—— Taraxaci	—— Dandelion	G.
♃	—— Auricula	—— narrow-leaved	G.
♃	—— paludofum	—— marſh	G.
♃ 597	—— Pilofella	L. —— Moufe-ear	G.
♃ 598	—— dubium	—— creeping	G.
♃ 599	—— murorum	—— Golden Lungwort	G.
♃ 600	—— fubaudum	L. —— ſhrubby	G.
♃ 601	——— umbellatum	L. —— buſhy	G.
☉ 602	Hyoſeris minima	L. Swine's-Succory leaſt	G.
♃ 603	Hordeum murinum	L. Barley-graſs wall	Gr.
♃ 604	——— ſylvaticum	—— wood	Gr.
♃ 605	—— pratenſe	L. —— meadow	Gr.
♃ 606	Holcus mollis	L. Soft-graſs wood	Gr.
♃ 607	—— lanatus	L. —— meadow	Gr.
♃ 608	Hottonia paluſtris	L. Water-violet	B.
♂ 609	Hyofcyamus niger	L. Henbane	G.
♃ 610	Herniaria glabra	Rupture-wort fmooth	W.
♃ 611	—— hirſuta	—— hairy	W.
♃ 612	Heraclium Sphon-dylium	L. Cow-parſnep	G.
♃ 613	Hedyfarum Ono-brychis	L. Saint-foin	G.
♃ 614	Hypericum An-droſæmum	L. Tutſan	G.
♃ 615	—— humifufum	L. St. John's-wort trailing	W.
♃ 616	—— pulchrum	L. —— fmall upright	G.
♃ 617	—— perforatum	L. —— common	G.
♃ 618	—— hirfutum	L. —— hairy	G.
♃ 619	—— montanum	L. —— mountain	G.
♃ 620	—— quadrangulum	L. —— fquare-ſtalked	G.
♃ 621	—— elodes	L. —— marſh	B.
♃	Iſoetes lacuſtris	Quillwort	B.
☉ 622	Iberis amara	Candy-tuft bitter	G.
♃ 623	Juncus acutus	Ruſh ſea	B.
♃ 624	—— conglomeratus	L. —— cluſtered	B.
♃ 625	—— effufus	L. —— foft	B.
♃ 626	—— ſquarrofus	L. —— goofe	B.
♃ 627	—— articulatus	L. —— jointed	B.

♃ 628 Juncus *compreſſus*	L. Ruſh *flat-ſtalked*	B.
♃ 629 —— *viviparus*	L. —— *viviparous*	B.
♃ 630 —— bulboſus	L. —— bulbous	B.
☉ 631 —— bufonius	L. —— toad	B.
♃ 632 —— *glaucus*	L. —— *glaucous*	B.
♃ —— ſpicatus	L. —— ſpiked	B.
♃ 633 Iris Pſeudacorus	L. Flag yellow	G.
♃ 634 —— fœtidiſſima	—— ſtinking	G.
☉ 635 Impatiens noli me tangere	Touch me not	Wd.
☉ 636 Limoſella aquatica	Mudwort	B.
♃ 637 Lavatera arborea	Tree Mallow	G.
♃ Lathyrus paluſtris	Vetch marſh	B.
☉ —— hirſutus	—— hairy	G.
♃ 638 —— pratenſis	L. —— meadow	G.
♃ 639 —— latifolius	L. { Everlaſting Pea broad leaved	G.
♃ 640 —— ſylveſtris	—— narrow-leaved	G.
♃ Lobelia Dortmanna	Gladiole water	B.
♃ —— urens	L. Lobelia ſtinking	G.
♃ 641 Lycopus europæus	L. Water-horehound	G.
☉ 642 Lolium temulentum	L. Darnel	Gr.
♃ 643 Lyſimachia thyrſiflora	Looſeſtrife tufted	B.
♃ 644 —— Nummularia	L. Moneywort	G.
♄ 645 Lonicera Pericly- menum	L. Honeyſuckle	Wd.
♃ 646 Liguſticum Sco- ticum	Lovage Scotiſh	W.
♃ * —— Cornubienſe	—— Corniſh	
☉ 647 Linum uſitatiſſimum	Flax common	G.
♃ 648 —— perenne	—— perennial	G.
♂ 649 —— anguſtifolium	L. —— narrow-leaved	G.
☉ 650 —— Radiola	L. —— leaſt	W.
☉ 651 —— catharticum	L. —— purging	W.
♃ 652 Lythrum Salicaria	{ Looſe-ſtrife purple- ſpiked	G.
☉ 653 —— hyſſopifolium	—— hyſſop-leaved	B.
♃ 654 Lepidium latifolium	Dittander broad-leaved	G.

☉ 655	Lepidium ruderale	L. { Dittander narrow-leaved	G.
♃ 656	Lotus corniculatus	L. Birds-foot Trefoil	W.
☉ 657	Lactuca virosa	L. Lettuce wild	G.
☉	—— ſcariola	L. —— prickly	G.
♃ 658	—— ſaligna	L. —— leaſt	G.
♃ 659	Leontodon hiſpidum }	L. Dandelion rough	G.
☉ 660	Lapſana communis	L. Nipplewort	G.
♃ 661	Littorella lacuſtris	Littorella ſmall	B.
☉	Milium lendigerum	Millet-graſs corn	Gr.
♃ 662	—— effuſum	—— wood	Gr.
♃	Monotropa Hypopithys }	Hypopitys yellow	
♃	Mentha ſativa	Mint curled	B.
♃ 663	—— viridis	—— ſpear	B.
♃ 664	—— villoſa	—— hoary	B.
♃ 665	—— ſylveſtris	—— horſe	B.
♃ 666	—— rotundifolia	—— round-leaved	B.
♃ 667	—— piperitis	—— pepper	B.
♃ 668	—— aquatica	—— water	B.
♃ 669	—— gentilis	—— red	B.
♃ 670	—— arvenſis	L. —— field	B.
♃ 671	—— Pulegium	Pennyroyal	B.
☉	Melampyrum criſtatum }	Cow-wheat creſted	Wd.
☉	—— arvenſe	—— corn	Wd.
☉	—— pratenſe	—— meadow	Wd.
☉	—— ſylvaticum	—— wood	Wd.
♂	Malva parviflora	Mallow ſmall-flowered	G.
♃ 672	—— ſylveſtris	—— common	G.
♃ 673	—— rotundifolia	—— dwarf	G.
♃ 674*	—— Alcea	L. —— Vervain	G.
♃ 675	—— moſchata	L. —— muſk	W.
♃ 676	Myriophyllum ſpicatum }	L. Water-milfoil ſpiked	B.
♃ 677	—— verticillatum	L. —— whirled	B.
♃ 678	Menyanthes nymphoides }	Water-lily fringed	B.

♃ 679	Melissa Calamintha	Calamint mountain	G.
♃ 680	Medicago sativa	Lucern	G.
♃ 681	—— falcata	Medick yellow	G.
⊙ 682	—— polymorpha	L. Claver	G.
⊙ 683	Matricaria Ca- momilla	L. Camomile corn	G.
⊙ 684	Mercurialis annua	L. Mercury French	G.
♃ 685	Nardus stricta	L. Mat-grafs	Gr.
♃ 686	Nymphea lutea	L. Water-lily yellow	B.
♃ 687	—— alba	L. —— white	B.
♃ 688	Nepeta Cataria	L. Catmint	G.
♃	Ophrys monorchis	Orchis mufk	W.
♃ 689	Oenanthe fistulosa	L. { Water-dropwort com- mon	B.
♃ 690	—— crocata	L. —— hemlock	B.
♃ 691	Origanum vulgare	L. Marjoram wild	G.
♂ 692	Onopordum A- canthium	L. Cotton-thistle	G.
♃ 693	Orchis bifolia	L. Orchis butterfly	W.
♃ 694	—— pyramidalis	—— pyramidal	W.
♃ 695	—— maculata	L. —— late-fpotted	W.
♃ 696	—— conopfea	—— fweet-fcented	W.
⊙ 697	Phleum arenarium	Cat's-tail grafs fea	Gr.
♃ 698	—— pratenfe	L. —— meadow	Gr.
♃ 699	—— nodofum	L. —— bulbous	Gr.
⊙ 600	—— paniculatum	—— panicled	Gr.
♃ 701	Poa compreffa	L. { Meadow-grafs flat- ftalked	Gr.
♃	—— alpina	—— alpine	Gr.
♃ 702	—— criftata	L. —— crefted	Gr.
♃ 703	—— maritima	—— fea	Gr.
♃	Potamogeton lucens	L. Pondweed fhining	B.
♃	—— compreffum	L. —— flat	B.
♃	—— gramineum	L. —— graffy	B.
♃	—— pectinatum	L. —— fine	B.
♃	—— fetaceum	—— fetaceous	B.
♃	—— marinum	—— fea	B.
♃	—— pufillum	—— fmall	B.
♃ 704	—— natans	L. —— broad-leaved	B.

♃	705 Potamogeton perfoliatum	L. Pondweed perfoliate	B.
♃	706 ——— crifpum	L. ——— curled	B.
♃	707 ——— denfum	L. ——— dichotomous	B.
⊙	708 Phalaris canarienfis	Canary-grafs common	Gr.
♃	709 ——— arundinacea	L. ——— reed	Gr.
⊙	——— *phleoides*	———— *cat's-tail*	Gr.
⊙	710 Polycarpon tetraphyllum	Polycarpon four-leaved	W.
♃	711 Plantago major	L. Plantain great	G.
♃	712 ——— media	L. ——— hoary	G.
♃	713 ——— coronopus	L. ——— buckfhorn	G.
♃	714 ——— maritima	——— fea	G.
♃	715 Polemonium cæruleum	Jacob's Ladder	G.
♃	716 Phyteuma orbicularis	Rampion-horned	G.
♂	717 Phellandrium aquaticum	L. Water-hemlock fine-leaved	B.
♂	718 Paftinaca fylveftris	L. Parfnep wild	G.
♃	719 Pimpinella major	Burnet-faxifrage large	G.
♃	720 ——— faxifraga	L. ——— fmall	G.
⊙	721 Peplis Portula	L. Purflane water	B.
♃	722 Pyrola rotundifolia	Winter-green round-leaved	W.
⊙	723 Prunella vulgaris	L. Self heal	G.
⊙	724 Pedicularis paluftris	L. Loufewort marfh	B.
♃	725 Polygala vulgaris	L. Milkwort	W.
♃	Pifum maritimum	Pea fea	G.
♃	726 Prenanthes muralis	L. Wild Lettuce ivy-leaved	G.
⊙	727 Picris echioides	L. Ox-tongue	G.
♃	——— hieracioides	Hawkweed rough	G.
♃	728 Poterium Sanguiforba	Burnet great	G.
♃	Panicum Dactylon	Panic-grafs creeping	Gr.
♃	729 Parietaria officinalis	L. Pellitory of the Wall	G.
♃	Rubia peregrina	Madder wild	G.
♃	Ruppia maritima	Ruppia fea	B.

♃ 730	Rumex maritimus	L. Dock-small water	G.
♃ 731	—— sanguineus	L. —— bloody	G.
♃ 732	—— crispus	L. —— curled	G.
♃ 733	—— pulcher	L. —— fiddle	G.
♃ 734	—— Hydrolapathum	L. —— Great Water	G.
♃ 735	—— obtusifolium	L. —— broad-leaved	G.
♃ 736	—— acutus	L. —— sharp-pointed	G.
♃ 737	—— digynus	Sorrel mountain	G.
♃ 738	Rubus idæus	Raspberry	Tr.
♃ 739	—— cæsius	L. Dewberry	G.
♃ 740	—— fruticosus	L. Bramble common	G.
♃ 741	Ranunculus Flammula	L. Spearwort small	B.
♃ 742	—— Lingua	L. —— great	B.
♃ 743	—— repens	L. Crowfoot creeping	G.
☉ 744	—— *hirsutus*	L. —— *hairy*	G.
♃	Scirpus romanus	Club-rush single-headed	B.
♃ 745	—— setaceus	L. —— least	B.
♃ 746	—— palustris	L. —— marsh	B.
♃ 747	—— lacustris	L. Bull rush	B.
♃ 748	—— mucronatus	L. Club-rush pointed	B.
♃ 749	—— maritimus	L. —— sea	B.
♃ 750	—— sylvaticus	L. —— wood	B.
♃	—— pauciflorus	—— few-flowered	B.
☉	Salsola Kali	Glasswort prickly	B.
♃	Saxifraga nivalis	Saxifrage mountain	W.
♃ 751	—— autumnalis	—— autumnal	W.
♃ 752	—— cespitosa	—— matted	W.
☉ 753	Silene anglica	L. Catchfly English	W.
☉	*Spergula pentandra*	*Spurry pentandrous*	W.
♃ 754	—— saginoides	L. —— pearlwort	W.
♃ 755	—— nodosa	L. —— knotted	W.
☉ 756	—— arvensis	L. —— common	W.
☉	Subularia aquatica	Awlwort	B.
☉	Sonchus alpinus	Sow-thistle alpine	G.
♃ 757	—— arvensis	L. —— corn	G.
♃ 758	—— palustris	L. —— marsh	G.
♃	Serratula alpina	Saw-wort mountain	G.
♃ 759	—— tinctoria	L. —— common	G.

♃ 760 Sagittaria sagittifolia	L. Arrow-head	B.
♃ *Satyrium fuscum*	*Satyrion brown*	W.
♃ 761 —— viride	—— green	W.
♃ 762 Senecio Jacobæa	L. Ragwort common	G.
♃ 763 —— aquatica	L. —— marsh	G.
♃ Santolina maritima	Santolina sea	G.
☉ 764 Salicornia herbacea	Glasswort jointed	B.
♃ 765 ——— *fruticosa*	——— *shrubby*	B.
♃ 766 Schænus Mariscus	Bogrush prickly	B.
♃ 767 ——— nigricans	—— black headed	B.
♃ 768 ——— compressus	—— compressed	B.
☉ 769 ——— albus	L. —— white	B.
☉ ——— *fuscus*	—— *brown*	B.
♃ 770 Stipa pinnata	Feather-grass	Gr.
♃ 771 Scabiosa arvensis	L. Scabious field	G.
♃ 772 ——— columbaria	—— mountain	G.
♃ 773 Sanguisorba officinalis	Burnet great	G.
♃ 774 Solanum Dulcamara	L. Nightshade woody	G.
♃ Swertia perennis	Marsh Gentian	B.
♃ 775 Selinum palustre	Selinum marsh	B.
♃ 776 Sison segetum	L. Honewort corn	G.
☉ 777 —— inundatum	L. —— water	B.
♃ 778 Sambucus Ebulus	L. Dwarf Elder	G.
♃ 779 Statice Armeria	Thrift	G.
♃ 780 ——— reticulata	Sea Lavender matted	B.
♃ 781 ——— Limonium	——— common	B.
♃ 782 Saponaria officinalis	L. Sopewort	G.
♃ 783 Sedum Telephium	L. Orpine	W.
♃ 784 —— reflexum	L. { Stone-crop sharp-pointed }	W.
♃ 785 —— rupestre	—— rock	W.
♃ 786 —— album	L. —— white	W.
♃ 787 —— acre	L. —— biting	W.
♃ 788 —— sexangulare	L. —— insipid	W.
♃ 789 —— Anglicum	—— English	W.

P

♃ 790	Sedum dafyphyl-lum	L. Stone-crop thick-leaved	W.
♂ 791	—— villofum	——— bog	B.
♃ 792	Sempervivum tectorum	L. Houfe-leck	W.
♃ 793	Spiræa Filipendula	L. Dropwort	G.
♃ 794	Stratiotes aloides	Fresh-water Soldier	B.
♃ 795	Schrophularia nodofa	L. Figwort knobby	G.
♃ 796	—— aquatica	L. —— Water Betony	G.
♃ 797	—— Scorodonia	—— Balm leaved	G.
♃ 798	Sibthorpia europæa	Baftard Moneywort	W.
♃ 799	Sinapis nigra	L. Muftard black	G.
☉ 800	—— alba	L. —— white	G.
☉ 801	Solidago cambrica	Golden-rod Welch	G.
♃	Serapias longifolia	Helleborine long-leaved	W'd
♃	—— grandiflora	—— great-flowered	W'd.
♃ 802	—— paluftris	—— marfh	W'd.
♃	—— latifolia	—— broad-leaved	W'd.
♃	—— anguftifolia	—— narrow-leaved	W'd.
♃	—— purpurafcens	—— red	W'd.
♃ 803	Sparganium erec-tum	L. Bur-red common	B.
♃ 804	—— natans	—— fmall	B.
♃	Triticum junceum	Wheat-grafs rufhy	Gr.
♃ 805	——— repens	L. Couch-grafs garden	Gr.
♂	Thlafpi hirtum	{ Mithridate-muftard rough	G.
♃ 806	Triglochin paluftre	L. { Arrow-headed-grafs marfh	B.
♃ 807	——— maritimum	L. ——— fea	B.
♃ 808	Thalictrum alpinum	Meadow-rue mountain	W.
♃ 809	——— minus	—— fmall	G.
♃ 810	——— flavum	L. —— common	G.
♃ 811	Teucrium Scordium	Scordium	B.
♃ 812	—— Chamædrys	Germander	G.
☉ 813	—— Chamæpitys	L. Ground pine	G.
♃ 814	—— Scorodonia	L. Wood Sage	G.

♃	815	Thymus Serpyllum	L. Thyme wild	W.
♃	816	—— Acinos	L. —— basil	W.
♃	817	Trifolium repens	L. Clover Dutch	G.
♃	818	—— ochroleucon	L. —— yellow	G.
♃	819	—— alpestre	L. —— alpine	G.
♃	820	—— maritimum	Trefoil sea	W.
☉	821	—— arvense	L. —— hare's-foot	W.
☉	822	—— striatum	L. —— striated	W.
☉	823	—— glomeratum	L. —— round-headed	W.
♃	824	—— fragiferum	L. —— strawberry	G.
☉	825	—— agrarium	L. —— hop	W.
☉	826	—— procumbens	L. —— procumbent	W.
☉	827	—— filiforme	L. —— least	W.
♃	828	Tanacetum vulgare	L. Tansy common	G.
♃	829	Typha latifolia	L. Cat's-tail large	B.
♃	830	—— angustifolia	L. —— small	B.
♃	831	Tormentilla offi-cinalis	L. Tormentil common	G.
♃	832	var. reptans	L. —— creeping	G.
♃		Utricularia vulgaris	L. Hooded-Milfoil common	B.
♃		—— minor	—— small	B.
☉	833	Urtica pilulifera	Nettle Roman	G.
☉	834	—— urens	L. —— small	G.
☉	835	—— dioica	L. —— common stinging	G.
♂	836	Verbascum Thapsus	L. Mullein tall	G.
♂	837	—— Lychnitis	L. —— Kentish	G.
♂	838	—— pulverulentum	—— Norfolk	G.
♂	839	—— Blattaria	L. —— moth	G.
♃	840	—— nigrum	L. —— black	G.
♃		Vicia hybrida	Vetch bastard	G.
♃		—— bithynica	—— rough	G.
♃	841	Veronica spicata	Speedwell spiked	G.
♃	842	—— hybrida	—— Welch	G.
♃	843	—— officinalis	L. —— male	G.
♃	844*	—— fruticulosa	—— shrubby	G.
♃	845	—— Becabunga	L. Brooklime	B.
♃	846	Valeriana officinalis	L. Valerian great	G.
♃	847	Vicia sylvatica	Vetch wood	G.

♃ 848 Vicia lutea	Vetch yellow	G.
♃ 849 —— fepium	L. —— bufh	G.
♃ 850 —— Cracca	L. —— tufted	G.
♃ 851 Verbena officinalis	L. Vervain	G.
☉ 852 Xanthium Stru- marium	Burdock leffer	G.
♃ Zoftera marina	Grafs-wrack	B.
♃ Zannichellia pa- luftris	L. Zannichellia	B.

A U G U S T.

The funny wall
Prefents the downy Peach, the fhining Plum,
The ruddy fragrant Nectarine, and dark,
Beneath his ample leaf, the lufcious Fig;
The Vine too, here, her curling tendrils fhoots,
Hangs out her clufters, glowing, to the fouth,
And fcarcely wifhes for a warmer fky.

Thomfon.

☉ 853 Antirrhinum E- latine	L. Fluellin fharp-pointed	W.
☉ 854 —— fpurium	L. —— round-leaved	W.
☉ 855 —— Orontium	L. Snapdragon fmall	W.
♃ 856 Arundo Calama groftis	L. Reed-grafs wood	Gr.
♂ 857 Angelica fylveftris	L. Angelica wild	B.
♃ 858 Allium carinatum	Garlic mountain	B.
♃ 859 —— vineale	L. —— crow	B.
♃ Anthericum ca- lyculatum	Afphodel Scottifh	B.

♃ 860	Agrimonia Eupatoria	L. Agrimony	G.
♃	Artemisia campestris	Wormwood field	G.
♃ 861	—— maritima	—— sea	G.
♃ 862	—— Abfinthium	L. —— common	G.
♃ 863	—— cærulefcens	L. —— blueifh	G.
♃ 864	—— vulgaris	Mugwort	G.
♃ 865	After Tripolium	Starwort sea	B.
♃ 866	Anthemis nobilis	L. Camomile common	G.
♃ 867	Achillea Ptarmica	L. Sneefewort	G.
☉ 868	Amaranthus Blitum	L. Amaranthus blite	G.
♃ 869	Atriplex portulacoides	Orach purflane-leaved	G.
☉ 870	—— laciniata	—— sea	G.
☉ 871	—— haftata	L. —— fpear	G.
☉ 872	—— patula	L. —— fpreading	G.
☉	—— ferrata	—— oak-leaved	G.
☉ 873	—— littoralis	—— grafs leaved	G.
☉	—— pedunculata	—— peduncled	G.
♃ 874	Adianthum Capillus	Maiden-hair true	W.
♃ 875	Afplenium Adiantum nigrum	L. —— black	W.
♃ 876	—— Ruta muraria	L. —— wall rue	W.
♃	—— marinum	—— sea	W.
♃	—— lanceolatum	—— fpear	W.
♃	—— viride	—— green	W.
♃ 877	—— Trichomanes	L. —— common	W.
♃ 878	—— Ceterach	Spleenwort	W.
♃ 879	—— Scolopendrium	Hart's-tongue	W.
♃ 880	Acroftichum feptentrionale	Acroftichum forked	W.
	—— ilvenfe	—— hairy	W.
☉	* Bromus fquarrofus	Brome-grafs field	Gr.
☉	Bartfia alpina	Bartfia alpine	W.
☉	—— vifcofa	—— clammy	W.
	881 Ballota nigra	L. Horehound ftinking	G.

☉ 882	Bidens cernua	L. Water-hemp-agrimony nodding	B.
☉ 883	—— tripartita	L. —— trifid	B.
♃ 884	Ceratophyllum demerfum	L. Horned-Pondweed	B.
♃ 885	Cyperus longus	Cyperus fweet	B.
♃	Carex incurva	Carex curved	B.
☉ 886	Cynofurus echinatus	Dog's-tail grafs prickly	Gr.
♃ 887	Convolvulus fepium	L. Bindweed large	G.
♃ 888	Campanula rotundifolia	L. Bell flower round-leaved	W.
☉ 889	Chenopodium urbicum	L. Goofe-foot upright	G.
♃ 890	—— Bonus Henricus	L. —— Good King Henry	G.
☉ 891	—— rubrum	L. —— red	G.
☉ 892	—— murale	L. —— nettle-leaved	G.
☉ 893	—— hybridum	L. —— thornapple-leaved	G.
☉ 894	—— album	L. —— common	G.
☉ 895	—— viride	L. —— red-jointed	G.
☉ 896	—— glaucum	L. —— oak-leaved	G.
☉ 897	—— maritimum	—— fea	G.
♃ 898	Ceraftium aquaticum	L. Moufe-ear Chickweed marfh	B.
♃ 899	Clematis Vitalba	L. Traveller's joy	G.
♃ 900	Conyza fquarrofa	L. Plowman's Spikenard	G.
♃ 901	Centaurea nigra	L. Knapweed common	G.
♃ 902	—— laciniata	L. —— jagged	G.
♃ 903	—— Scabiofa	L. —— great	G.
♂ 904	—— Calcitrapa	L. Star-thiftle	G.
♃	Dactylis ftricta	Cock's-foot-grafs water	Gr.
♂ 905	Dipfacus fylveftris	L. Teazel wild	G.
♂	—— fullonum	—— manured	G.
♂ 906	—— pilofus	L. —— fmall	G.
☉ 907	Datura Stramonium	L. Thorn-apple	G.
♃ 908	Erica vulgaris	L. Heath common	Wd.
♃ 909	—— cinerea	L. —— fine-leaved	Wd.
♃ 910	—— tetralix	L. —— crofs-leaved	Wd.
♃ 911	—— multiflora	—— many-flowered	Wd.

☉ 912	Euphorbia Peplus	L. Spurge garden	G.
☉ 913	—— Peplis	—— purple	G.
☉ 914	—— exigua	L. —— petty	G.
♃ 915	—— portlandica	—— Portland	G.
☉ 916	—— Heliofcopia	L. —— fun	G.
☉ 917	—— platyphyllos	L. —— broad-leaved	G.
♃ 918	—— Hyberna	—— Irifh	W.
☉ 919	Euphrafia Odontites	L. Eye-bright red	W.
☉ 920	—— officinalis	L. —— common	W.
☉ 921	Erigeron canadenfe	L. Erigeron Canada	G.
♃ 922	Equifetum hyemale	Shave-grafs	B.
♃	* Feftuca rubra	Fefcue-grafs red	Gr.
♃	—— Cambrica	—— Welch	Gr.
♃	Galium pufillum	Ladies-bedftraw fmall	W.
☉ 923	—— Anglicum	—— Englifh	B.
♃ 924	—— boreale	—— crofs-leaved	G.
☉ 925	Gentiana Amarella	Gentian autumnal	W.
☉	—— campeftris	—— field	W.
☉	—— filiformis	—— leaft	W.
☉ 926	Gnaphalium montanum }	L. Cudweed leaft	W.
☉	—— gallicum	—— corn	W.
♃ 927	—— margaritaceum	—— everlafting	G.
☉	—— fupinum	—— dwarf	W.
♃ 928	—— fylvaticum	L. —— wood	W.
☉ 929	—— uliginofum	L. —— black-headed	W.
☉ 930	—— germanicum	L. —— common	W.
♃	Juncus filiformis	Rufh leaft	B.
♃	—— biglumis	—— two-glum'd	B.
☉ 931	Illecebrum verti cillatum }	Illecebrum marfh	B.
♃ 932	Inula Helenium	Elecampane	G.
♃ 933	—— dyfenterica	L. Fleabane common	G.
☉ 934	—— pulicaria	L. —— fmall	G.
♃ 935	—— crithmoides	Samphire golden	G.
♃ 936	Lithofpermum officinale }	L. Gromwel common	G.
♃ 937	Lyfimachia vulgaris }	L. Loofe-ftrife yellow	G.

24 938 Leonurus Cardiaca | L. Motherwort | G.
⊙ 939 Lemna minor | L. Duck-weed fmall | B.
⊙ 940 —— *gibba* | L. ——*gibbous* | B.
⊙ 941 —— polyrrhiza | L. —— large | B.
⊙ 942 —— trifulca | L. —— ivy-leaved | B.
24 943 Melica cærulea | L. Melic-grafs blue | Gr.
24 944 Marrubium vulgare | L. Horehound common | G.
24 945 Meliffa Nepeta | L. Catmint | G.
24 946 Matricaria Par-⎱thenium ⎰ | L. Feverfew | G.
⊙ 947 —— inodora | L. Camomile weak-fcented G.
948 Origanum Onites | Marjoram pot | G.
Ophrys corallorhiza | Orchis coral-rooted | W.
949 Ofmunda fpicant | L. Spleenwort rough | W.
950 —— crifpa | Fernftone | W.
951 —— regalis | Ofmund-royal | W.
952 Panicum viride | L. Panic-grafs green | Gr.
953 —— verticillatum | L. —— whirled | Gr.
954 —— fanguinale | L. —— cock's foot | Gr.
955 —— Crus galli | L. —— loofe | Gr.
956 Poa aquatica | L. Reed-grafs water | Gr.
957 Peucedanum Silaus | L. Saxifrage meadow | G.
958 Parnaffia paluftris | Grafs of Parnaffus | B.
959 Polygonum am-⎱phibium ⎰ | L. Perficaria amphibious | G.
960 —— Perficaria | L. —— mild | G.
961 —— *penfylvanicum* | L. —— *pale-flowered* | G.
962 —— Hydropiper | L. —— biting | G.
963 —— *minus* | L. —— *fmall* | G.
964 —— aviculare | L. Knot-grafs | G.
965 —— Convolvulus | L. Buck-wheat climbing | G.
966 —— Fagopyrum | L. —— common |
967 Polypodium Dry-⎱opteris ⎰ | Polypody branched | W.
—— fontanum | —— rock | W.
968 —— fragile | —— brittle | W.
—— lonchitis | —— rough | W.
24 —— phegopteris | —— wood | W.
24 969 —— aculeatum | L. —— prickly | W.

♃	Polypodium The-lypteris	Polypody marsh	W.
♃ 970	—— Filix fœmina	L. —— female	W.
♃ 971	—— Filix mas	L. —— male	W.
♃	—— lobatum	—— lobed	W.
♃ 972	—— criftatum	L. —— crefted	W.
♃	—— fragrans	—— fweet	W.
♃ 973	—— vulgare	L. —— common	W.
♃	—— rhæticum	—— ftone	W.
♃ 974	Pteris aquilina	L. Fern or Brakes	W.
♃	Schænus ferrrugineus	Bog-rufh ferrugineous	B.
♃	—— rufus	—— red	B.
♃ 975	Scirpus cefpitofus	L. Club-rufh fluted	B.
♃ 976	—— acicularis	L. —— fmooth	B.
♃ 977	—— fluitans	L. —— floating	B.
♃	—— holofchænus	—— round-headed	B.
♃ 978	Sifon verticillatum	Honewort whirled	G.
♂ 979	—— Amomum	L. Stone-parfley	G.
♃	Senecio paludofus	Ragwort marfh	B.
♃ 980	—— crucæfolius	L. —— hoary	G.
♃ 981	—— farracenicus	—— broad-leaved	G.
☉ 982	Solanum nigrum	L. Nightfhade garden	G.
♃ 983	Samolus Valerandi	L. Water-Pimpernel	B.
♃	Salfola fruticofa	Saltwort fhrubby	B.
♃ 984	Sium latifolium	L. Water-parfnep large	B.
♃ 985	—— anguftifolium	L. —— narrow-leaved	B.
♃ 986	—— nodiflorum	L. —— creeping	B.
♃ 987	Spiræa Ulmaria	L. Meadow-fweet	G.
♃ 988	Stachys fylvatica	L. Dead-nettle hedge	G.
♃ 989	—— paluftris	L. Clown's Wound-wort	G.
♃ 990	—— germanica	Bafe-horehound	G.
☉ 991	—— arvenfis	L. Stachys corn	G..
♃ 992	Scutellaria minor	L. { Hooded-willow-herb fmall	B.
♃ 993	—— galericulata	L. ———— common	B.
♃ 994	Solidago Virga aurea	L. Golden-rod common	G.

Q

♃ 995	Sagittaria fagitti-folia	L. Arrow-head	B.
♃	Trichomanes py-xidiferum	Trichomanes cup	
♃	—— Tunbrigenfe	—— Tunbridge	
♄ 996	Tilia europæa	L. Lime-tree	Tr.
♃ 997	Veronica fcutellata	L. Speedwell bog	B.
☉ 998	—— Anagallis a-quatica	L. —— water	B.

SEPTEMBER.

Ye Virgins come, for you their lateft fong
The woodlands raife ; the cluft'ring Nuts for you
The lover finds, amid the feeret fhade,
And where they burnifh on the topmoft bough,
With active vigour crufhes down the tree,
Or fhakes them ripe from the refigning hufk.

Thomfon.

♄ 999	Arbutus Unedo	Strawberry-tree common	Tr.
♃ 1000	Arundo Phrag-mites	Reed common	Gr.
♃ 1001	Colchicum au-tumnale	Meadow-faffron.	W.
♃ 1002	Crocus officinalis	Saffron true	W.
♃ 1003	Erica daboccii	Heath Irifh	Wd.
♄ 1004	Hedera Helix	Ivy	Tr.
♃ 1005	Leontodon au-tumnale	Dandelion autumnal	G.
♃ 1006	Ophrys fpiralis	Ladies-traces	W.
♃ 1007	Scabiofa Succifa	Devil's-bit	G.
♃ 1008	Scilla autumnalis	Squil autumnal	W.

O C T O B E R.

Thofe virgin leaves, of pureft vivid green,
Which charm'd 'ere yet they trembled on the trees,
Now chear the fober landfcape in decay :
The Lime firft-fading, and the golden Birch,
With bark of filver hue, the mofs-grown Oak,
Tenacious of its leaves of ruffet brown,
Th'enfanguin'd Dogwood, and a thoufand tints,
Which Flora, drefs'd in all her pride of bloom,
Could fcarcely equal, decorate the groves.
Be quick, ye artifts, 'ere a biting froft
Gives fcenes like thefe to fcatt'ring winds ; go, bid
In mimic hues deciduous beauties live.

N O V E M B E R.

                        Now the leaf
Inceffant ruftles from the mournful grove,
Oft' ftarting fuch as ftudious walk below,
And flowly circles through the waving air.
                        Thomfon.

D E C E M B E R.

No mark of vegetable life is feen,
    No bird to bird repeats his tuneful call,
Save the dark leaves of fome rude evergreen,
    Save the lone red-breaft on the mofs-grown wall.
            Q 2   *Scott.*

CONCLUSION.

Was every faultering tongue of man,
Almighty Father! filent in thy praife,
Thy works themfelves would raife a general voice,
Ev'n in the depths of folitary woods,
By human foot untrod ; proclaim thy pow'r,
And to the quire celeftial THEE refound,
Th'eternal Caufe, Support, and End of all.

Thomfon.

THE

ENGLISH NAMES

OF THE

BRITISH PLANTS,

ARRANGED ALPHABETICALLY.

R

S

A

C A T A L O G U E

O F

B O O K S,

I N

NATURAL HISTORY, MEDICINE,
AND AGRICULTURE,

DEPOSITED IN THE

LIBRARY OF THE GARDEN,

FOR THE USE OF SUBSCRIBERS.

ADANSON *(Franc.) Familles des Plantes, tom.* 2. 8*vo.*
1763.

AGRICULTURAL SOCIETY *(at Bath) Letters and Papers
on Agriculture, Planting, &c.* 8*vo.* 1780.

ALSTON *(Charles) Lectures on the Materia Medica,*
published by Dr. Hope, 2 *vol.* 4*to.* 1770.

ANDERSON. *Essays relating to Agriculture and rural
Affairs, vol.* 2. 8*vo.* 1775.

D'ASSO. *Synopsis Stirpium indigenarum Arragoniæ, et
Mantissa.* 8*vo.* 1779.

BATARRA *(A. J. Ant.) Fungorum Agri Arimensis Historia, cum Iconibus.* 4*to.* 1759.

BAUHINUS *(Casp.) Pinax. 4to. 1623.*
BERGIUS *(Petr. Jonas) Materia Medica, tom.* 2. *8vo.*
1778.
BERKENHOUT *(John) Outlines of the Natural History*
of Great-Britain and Ireland. vol. 2d. 8vo. 1770.
——————————— *Clavis Anglica Linguæ Botanicæ.*
12mo. 1764.
BLAIR *(Patrick) Botanic Essays. 8vo. 1720.*
——————— *Pharmaco-Botanologia. 4to. 1733.*
BLACKSTONE *(J.) Specimen Botanicum. 8vo. 1746.*
BOERHAVE *(Herman.) Materia Medica. 12mo. 1755.*
BOTANICAL SOCIETY *(at Lichfield) System of Vegetables.*
8vo. Part 1. 2. 1782.
BRUNNICH *(M. Th.) Fundamenta Zoologiæ. 8vo.*
1772.
——————— *Entomologia. 8vo. 1760.*
BUTLER *(Ch.) History of Bees. 4to. 1634.*

CULLEN *(William) Lectures on the Materia Medica.*
4to. 1772.
CURTIS *(William) Flora Londinensis.* Folio. Vol. 1.
Plates coloured. 1777.
——————— *Illustration of Linnæus' Classes and*
Orders of Plants. With original Plates. 4to. Co-
loured. 1777.
——————— *Instructions for collecting and pre-*
serving Insects. 8vo. 1771.
——————— *Short History of the Brown-tail*
Moth. 4to. 1782.

DALE *(Sam.) Pharmacologia. 4to. 1737.*
DEERING *(C.) Catalogus Stirpium, &c. 8vo. 1738.*
DILLENIUS *(Jo. Jac.) Catalogus Plantarum circa*
Gissam nascentium. 12mo. 1719.
——————— *Hortus Elthamensis. Fol. 1774.*

'Evelyn *(John) Sylva, or Difcourfe on Foreft Trees, with Notes by Dr. A. Hunter. 4to. Plates. 1776.*
———————— *Kalendarium Hortenfe. 8vo. 1691.*
Edit. 8.

Fordyce *(Geo.) Elements of Agriculture and Vegetation. 8vo. 1771.*
Forster *(Joan. Rein.) Novæ fpecies Infectorum Centur. 1. 8vo. 1761.*
———————— *A Catalogue of Britifh Infects. 8vo. 1770.*
Fuchsius *(Leohn.) Icones Plantarum. 8vo. Imperfect.*

Gataker *(Thomas) Obfervations on the internal Ufe of Nightfhade. 8vo. 1757.*
Geoffroy *Hiftoire abregée des Infectes qui fe trouvent aux environs de Paris, avec Planches. Tom. 2. 4to. 1762.*
Goedartius *(Joh.) de Infectis Opera M. Lifter. 8vo. 1685.*
Grew *(Nehem.) Anatomy of Plants. Fol. 1682.*

Haller *(Albert) Opufcula. 8vo. 1749.*
Hales *(Stephen) Vegetable Staticks. 8vo. 1727.*
Hermannus *(Paul) Paradifus Batavus. 4to. 1648.*
———————— *Cynofura Materiæ Medicæ. 4to. 1710.*
Hudson *(Gul.) Flora Anglica. Edit. 1. 8vo. 1762.*
———————— *Flora Anglica. 8vo. Edit. 2. 1778.*

Jacob *(Ed.) Plantæ Faverfhamienfes. 8vo. 1777.*
Jacquin *(Nic. Jof.) Flora Auftriaca. Fol. Vol. 1, 2, 3, 4, 5. 1773.—1778.*
———————— *Mifcellanea Auftriaca. Vol. 1. 4to. 1778.*

T

JENKINSON *(James) Generic and Specific Descriptions of the British Plants from Linnæus.* 8vo. 1775.

Index Plantarum officinalium in Horto Chelsejano. 12mo. 1730.

KENT *(Nath.) Hints to Gentlemen of landed Property.* 8vo. 1775.

KONIG *(Eman.) Regnum Vegetabile.* 4to. 1696.

LEE *(James) Introduction to Botany.* 8vo. 3d Edit. 1776.

LESSER *(Monf.) Theologie des Insectes.* 8vo. 1742. par Lyonnet.

LETTSOM *(John Coakley) Naturalists and Travellers Companion.* 8vo. 1772.

LEWIS *(William) New Dispensatory.* 8vo. 1770.

LIGHTFOOT *(John) Flora Scotica.* 8vo. *Vol.* 2. 1777.

LINNÆUS *(Car.) Systema Vegetabilium, a Murray.* 8vo. 1774.

——————— *Mantissa Plantarum altera.* 8vo. 1771.
——————— *Spec. Plantarum. Edit.* 3. *Vol.* 2. 1764.
——————— *Genera Plantarum.* 8vo. 1764.
——————— *Bibliotheca Botanica.* 12mo. 1736.
——————— *Philosophia Botanica.* 8vo. 1763.
——————— *Systema Naturæ. Tom.* 3. 8vo. 1768.
——————— *Amœnitates Academicæ. Vol.* 1.-5, 6, 7. 1749-69.

——————— *Fauna Suecica.* 8vo. 1761.
——————— *Materia Medica. Edit. Schreber.* 8vo. 1762.

LISTER *(Mart.) Exercitatio Anatomica de Cochleis, &c.* 8vo. 1694.

LUDWIG *(Chrift. Gotlieb.) Institutiones Historico-physicæ regni Vegetabilis.* 8vo. 1757.

MALPIGHI *(Marcel.) Opera omnia. Vol.* 2. *4to.*
1687.

———————— *Anatome Plantarum. Fol.* 1675.

MATTHIOLUS *(Petr. Andr.) Commentaria in fex Libros
Diofcoridis de Medica Materia plurimis Iconibus il-
luftrata. Fol.* 1565.

MERIAN *(Mar. Sibil.) Erucarum Ortus Alimentum et
paradoxa Metamorphofis.* 4to. 1717.

MERRETT *(Chrift:) Pinax rerum naturalium Britannicarum.* 8vo. 1667.

MICHELI *(Petr. Ant.) Nova Plantarum Genera.* 4to.
Cum Iconibus. 1729.

MILLER *(John) Illuftration of the Sexual Syftem of Linnæus,* 1 *vol.* 8vo. 1779.

MILLER *(Phil.) Gardener's Dictionary abridged.* 6th Edit.
4to. 1771.

———————— *Gardener's Kalendar. Edit.* 16. 8vo.
1765.

MILNE *(Colin) Inftitutes of Botany.* 2 *Parts.* 4to.
1771.

———————— *Botanical Dictionary.* 8vo. 1770.

MOUFFET *(Thom) Theatrum Infectorum. Fol. cum
Iconibus.* 1634.

NEALE *(Adam) Catalogue of the Plants in Mr. Black-
burn's Garden at Orford.* 1 *Vol.* 8vo. 1779.

NECKER *(N. J. de) Methodus Mufcorum.* 8vo. 1771.

NEWMAN *(Cafp.) Chemical Works, by Lewis.* 2d Edit.
Vol. 2. 8vo.

Nomenclator Botanicus Lipfiæ. 8vo. 1772.

OEDER *(Geo. Chrift.) Enumeratio Plantarum Floræ
Danicæ, Pars prima.* 8vo. 1770.

———————— *Elementa Botanicæ.* 8vo. 1764.

T 2

PARKINSON *(John) Theatre of Plants. Folio.* 1640.
Pharmacopœia Londinensis. 8vo. 1763.
Pharmacopœia Collegii regii Medicorum Edinburgensis.
8vo. 1774.
PLAT *(Hugh) Garden of Eden.* 12mo. 1653.
PULTNEY *(Richard) General View of the Writings of
Linnæus.* 8vo. 1781.

RAJUS *(Joh.) Synopsis Stirpium Britannicarum.* 8vo.
Edit. 3d. 1724.
——————— *Catalogus Plantarum circa Cantabri-
giam.* 12mo. 1660.
REDI *(Franciscus) Opuscula de Insectis, &c.* Tom. 2.
12mo. 1636.
ROBSON *(Stephen) British Flora,* 1 Vol. 8vo. 1777.
ROSE *(Hugh) Elements of Botany,* 1 Vol. 8vo. 1775.
ROTHERAM *(John) Caroli a Linnè Termini Botanici.*
12mo. 1779.
ROTTBOLL *(Christ. Friis.) Novarum Plantarum De-
scriptiones et Icones. Fol.* 1778.
RUPPIUS *(Hen. Bernh.) Flora Jenensis. Edit. Haller.*
8vo. 1745.
RUTTY *(Joh.) Materia Medica illustrata.* 4to. 1775.

SCHEUCHZER *(Joh.) Agrostographia,* 1 Vol. 4to. *Edit.
Heller.* 1775.
SCHREBER *(Joh. Christ. Dan.) Beschreibung der Gräser.
Fol.* 1769.
——————— *Observationes de Phasco, cum
Iconibus.* 4to. 1770.
——————— *Lithographia Hallensis.* 8vo.
1759.
SCOPOLI *(Joh. Ant.) Flora Carniolica.* Tom. 2. 8vo.
1772.

Scopoli *(Joh. Ant.)* Entomologia Carniolica. 8vo. 1763.

———————— Annus Historico Naturalis, 8vo. 1769.

Society of Physicians *(in London)* Medical Observations and Enquiries. *Vol.* 5th. 1776.

Stillingfleet *(Benjamin)* Miscellaneous Tracts. 2d Edit. 8vo. 1762.

Swammerdam *(Jean)* Histoire Generale des Insectes. 4to. 1685.

Tournefort *(Jos. Pitton.)* Institutiones Rei Herbariæ. Tom. 3. 4to. 1700.

———————— Histoire des Plantes qui Naissent aux Environs de Paris. 12mo. 1648.

Vaillant *(Sebast.)* Botanicon Parisiense. Fol. 1727.

———————— De Structura Florum. 4to. 1727.

Van Royen *(Adrian.)* De Anatome et Oeconomia Plantarum. 4to. 1728. Diss. inaug.

Walcott *(John)* Flora Britannica indigena. 14 Numbers. 8vo. 1778.

Warner *(Richard)* Plantæ Woodfordienses. 8vo. 1771.

Weis *(Frid. Gul.)* Plantæ Cryptogamicæ. 8vo. 1770.

Wepferus *(Joh. Jac.)* Cicutæ aquaticæ Historia et noxæ. 4to. 1679.

Willich *(Christ. Lud.)* De Plantis quibusdam Observationes. 8vo. 1762.

Weston *(Richard)* English Flora. 1 Vol. 8vo. 1775.

Withering *(William)* Botanical Arrangement of the British Vegetables. 2 Vol. 8vo. 1776.

Yeats *(Th. Pattinson)* Institutes of Entomology. 8vo. 1773.

To promote a Knowledge of the Plants of our own Country.

There is now publishing periodically in *Numbers*, price 5s. coloured, or 2s. 6d. plain,

The FLORA LONDINENSIS

A new and original Botanic Work, intended to comprehend all the *Plants* which grow wild in *Great-Britain*, beginning first with those which are found *in the Environs of London*.

By WILLIAM CURTIS.

This Work is of a *Folio* size, each Number contains *six Plates*, having in general *one Plant* on a Plate, with *six Pages* and sometimes more of Letter-press; the Plants are represented of their natural Size, and in their true Colours; in the Letter-press are given first the *Linnæan*, and *English Names*, next the *Synonyma* of the most interesting Authors, then follows a *minute Description* of the Plant in Latin and English, which is succeeded by an Account of the *History, Peculiarities, Oeconomy*, and *Uses* of the Plant, either selected from the most approved Authors on *Medicine, Agriculture, rural Oeconomy*, and *other Arts*; or furnished from the Observations of the Author and his Friends.

Published by the AUTHOR at his *Botanic Garden*, and *B. White, Fleetstreet*, where any of the Numbers from 1 to 46 may be had, either plain or coloured, as also the following:

www.ingramcontent.com/pod-product-compliance
Lightning Source LLC
Chambersburg PA
CBHW021817190326
41518CB00007B/637